HANDBOOK FOR NEW ACTORS IN SPACE

2ND EDITION, 2024

www.swfound.org

2nd Edition, 2024

ISBN 979-8-9882626-6-4

SECURE WORLD FOUNDATION

Secure World Foundation is a private operating foundation that promotes cooperative solutions for space sustainability and the peaceful uses of outer space. The Foundation acts as a research body, convener, and facilitator to promote key space security and other space sustainability-related topics and to examine their influence on governance and international development.

Edited by:

Dr. Peter Martinez, *Executive Director*

Lisa G. Croy, *Chief Operating Officer*

Victoria Samson, *Chief Director, Space Security and Stability*

Krystal Azelton, *Senior Director, Program Planning*

Ian A. Christensen, *Senior Director, Private Sector Programs*

Christopher D. Johnson, *Director, Legal Affairs and Space Law*

Robert Pemberton, *Director, Communications*

Tamara Tanso, *DC Operations Associate*

ACKNOWLEDGMENTS

The Secure World Foundation was assisted by many experts from governments, space agencies, private industry, and academia in the drafting of the first edition of this Handbook, its translations, and this second edition. This Handbook reflects the position and views of the Secure World Foundation and not those of the experts consulted. While all errors or omissions are entirely our fault, we are deeply grateful to the following individuals and organizations for their views and input.

Agencia Espacial Mexicana, Carlos Enrique Alvarado, Kahina Aoudia, P.J. Blount, Gérard Brachet, Dennis Burnett, Victoria Carter-Cortez, Sergio Camacho, Shenyan Chen, Richard DalBello, Laura Delgado López, Matt Duncan, Andrew D'Uva, Stephen Earle, Paul Frakes, Jonathan Garon, Bailey Geist, Michael Gleason, Andrea Harrington, Henry Hertzfeld, Talal Al Kaissi, T.S. Kelso, Karl Kensinger, Rich Leshner, Mike Lindsay, Peter Marquez, Tanja Masson-Zwaan, T.J. Mathieson, Nicolas Maubert, Steve Mirmina, Philippe Moreels, Monique Moury, Clay Mowry, Mark Mulholland, Yui Nakama, Nicole Nir, Gerald Oberst, Nobu Okada, Mazlan Othman, Xavier Pasco, Ivan Petitville, Kevin Pomfret, Daniel Porras, Myland Pride, Ruth Pritchard-Kelly, Rajeswari Rajagopalan, Ben Reed, Jay Santee, Satellite Associates, Franceska Schroeder, Michael K. Simpson, Kazuto Suzuki, Glenn Tallia, Gary Thatcher, Seth Walton, Guoyu Wang, Brian Weeden, Joshua Wolny, Greg Wyler, Ciro Arévalo Yepes, Yuan Yue, Jessica Young, and Luis Zea.

TABLE OF CONTENTS

CHAPTER ONE
The International
Framework for
Space Activities

LIST OF FIGURES

LIST OF TABLES

Peter Martinez, PhD
EXECUTIVE DIRECTOR
Secure World Foundation

Foreword

The Secure World Foundation is proud to present the second edition of this *Handbook for New Actors in Space*.

By whatever metric one chooses, the global space arena is expanding at a breathtaking pace. When the first edition of this Handbook was published in 2017, there were some 1,700 active satellites in space and the space economy was valued at around $383.5 billion. Now, seven years later, in mid-2024, there are more than 10,200 active satellites in orbit and the global space economy has grown to more than $630 billion. Much of this growth is fueled by the commercial space sector and entrepreneurs who have access to more abundant capital, access to more commercial sources of flight-proven off-the-shelf space components, and more opportunities to access space at a lower launch cost per kilogram than was the case seven years ago. These trends are supporting the development of new kinds of space applications for societal benefit and new capabilities for space logistics, such as on-orbit refueling and servicing, and in-space assembly and manufacturing. These on-orbit capabilities will be foundational for the continued growth and development of the global space economy. At the same time, the greater pace of space activities is placing more responsibilities for regulation and oversight on the governments of countries involved in the space arena.

Space is a domain beyond national jurisdiction, but which borders all nations on Earth. Regardless of whether one is a commercial actor, an academic actor, or a state actor, it is important to have a basic knowledge and understanding of how space activities are governed and regulated at national and international level. It is also important to be cognizant of norms and best practices for safe and sustainable space operations. The goal of this Handbook is to provide new actors in space with a broad overview of the fundamental principles, laws, norms, and best practices for the safe and responsible conduct of peaceful activities in outer space. It is intended to reach two broad categories of new actors: firstly, those working in national governments beginning to develop their national space policies and regulations, and secondly, those working in expanding or start-up companies, investment and insurance firms, universities, and other non-governmental entities taking new steps into the space arena.

As in the first edition of the Handbook, this updated second edition is structured in three chapters. The first chapter focuses on what new actors should know about the international legal framework for space activities, the second chapter focuses on national policies and the regulation of space activities, and the third chapter deals with the safe and responsible conduct of space activities, through all the phases of a space project.

The content of the Handbook has been revised and updated extensively to reflect developments in the space arena since the publication of first edition. These updates include new sections on: the UN COPUOS Guidelines for the Long-Term Sustainability of Outer Space Activities (LTS Guidelines); the UN Online Index of Objects Launched Into Outer Space; Dark and Quiet Skies; Orbital Carrying Capacity; Potential Atmospheric Effects from Launches and Re-entries; and Sustainable Practices for Cislunar Space and Lunar Surface Operations; In-Space Servicing, Assembly, and Manufacturing; and Commercial Human Spaceflight Safety. Several new case studies have also been added.

Despite the rapid pace of development since the publication of the first edition of the Handbook, the basic fundamentals of space governance and safe and sustainable space operations haven't changed all that much. What has changed is that the space arena is becoming much more crowded and complex, with many new actors entering from other domains. These new actors bring with them a wealth of expertise into the space arena, but they may lack knowledge of the basic "rules of the road" for space actors. We hope that this Handbook will provide a helpful introduction to space governance and the safe and sustainable conduct of space activities for those new actors making their first forays into the space arena. ■

Peter Martinez is the Executive Director of the Secure World Foundation. He has extensive experience in multilateral space diplomacy, space policy formulation and space regulation. He also has extensive experience in capacity building in space science and technology and in workforce development. Prior to joining SWF, from 2011 - 2018 he chaired the United Nations Committee on the Peaceful Uses of Outer Space (UN COPUOS) Working Group on the Long-Term Sustainability of Outer Space Activities that negotiated a set of international consensus guidelines to promote the safety and sustainability of space operations.

HOW TO USE THIS BOOK

This Handbook is structured in three main chapters. Though meant to complement each other for a broad understanding of the entire scope of concern to new actors, certain chapters and sections will be of heightened interest to readers depending on their own expected space activity and the role that they will be playing in that activity.

CHAPTER ONE deals with the international legal and political order applicable to space activities, and gives an introduction to the most important and relevant topics in international space law and how they apply to states.

CHAPTER TWO discusses how national space policy and national regulation apply to space, beginning with rationales for developing space policy and discussing in particular how to broadcast goals internationally and give guidance domestically. The chapter also includes a discussion of the common aspects of national space legislation

CHAPTER THREE addresses responsible space operations and provides an overview of the process from pre-launch frequency selections and coordination to payload review, launch services agreements between launch providers and operators, and mission and post-mission concerns. More technical than Chapters One or Two, this final chapter explores the operational side of space activities.

Last, while textbooks on any of the various topics discussed in this book run into many hundreds of pages, this book aims to be both concise and readable. Rather than an exhaustive compendium of every facet and nuance of this incredibly rich field, this commentary is broad but contains only the most fundamental principles and topics with links to additional outside resources as appropriate.

Finally, readers should be mindful that the space arena is growing and evolving rapidly, so any statistics or other numbers listed in this Handbook should only be used with independent verification of the latest figures. Where possible, we've provided a link to resources and references for the reader to access. Any numbers quoted in this edition were correct as of May 2024 and updates to the online version after that date will be noted.

HANDBOOK FOR
NEW ACTORS
IN SPACE

CHAPTER ONE
The International Framework for Space Activities

CHAPTER ONE FOCUS

The focus of Chapter One is the international legal and regulatory framework, beginning with the rights and obligations of the Outer Space Treaty (OST) and the subsequent space treaties which expand and elaborate upon it, and especially the treaty's obligations in terms of international state responsibility and international registration of space objects. Relevant international organizations and multilateral fora are also included and discussed. International frequency management is then presented, as well as international principles relevant to remote sensing, international standards and standard-generating bodies in the space sector, and international export control measures. Discussion of state liability and the various avenues of dispute settlement then follows.

Various international environmental concerns are then explored, including protection of the Earth environment, back contamination of the Earth from space missions, nuclear power sources in space, space debris, and the protection of celestial bodies. To conclude the chapter, more advanced issues are explored, including the unresolved issues related to the lack of a legal definition of where outer space begins, the legal status and protections of humans in space, the effects of satellites in low Earth orbit on terrestrial activities, the use of space resources, and future governance issues.

This international framework for the conduct of space activities should be explored and understood by all governmental and non-governmental space actors so that they can be cognizant of the licensing and regulatory processes associated with space activities, and also as a general due-diligence exercise to ensure that they adhere to responsible and sustainable practices in space.

Simonetta Di Pippo

*Professor of Practice of Space
Economy & Director of the Space Economy Evolution Lab
(SEELab), SDA Bocconi; Director of the United Nations Office
for Outer Space Affairs (UNOOSA), 2014-2022*

INTRODUCTION

In 2017, the Secure World Foundation published the first version of the Handbook for New Actors in Space. The first version of the Handbook had two main audiences: governments new to space activities and private entities like commercial space startups. While governments and commercial companies have different aims and interests in conducting space activities, the SWF Handbook contains crucial information and insights for both.

Since the first edition of the Handbook came out, the diversity of actors looking upward to space has only increased. Likewise, the *range of activities* that new entrants to space are aiming to undertake has also only continued to grow and diversify. Meanwhile, the audiences that stand to benefit from the Handbook have also continued to expand. This is because, as Chapter 1 explains, the international legal and regulatory framework governing space activities is broadly applicable to all actors in the space domain—be it the government, private sector, the scientific community, academia, civil society, or otherwise—with Member States remain in charge of all the commitments they took in adhering to the Treaties and in approving the relevant guidelines.

Indeed, everyone interested in space activities, even if just how they impact other areas, should be aware of the international regime discussed in Chapter 1. Diplomats, legislators, regulators, as well as engineers, scientists, and principal investigators are just some of the types of stakeholders who should have a basic understanding of the international legal, regulatory, and even political context surrounding the regulation of humankind's activities in outer space. Space activities, in fact, cannot be conducted responsibly without knowledge on the full range of topics mentioned. To strive, this is the recipe.

In my role as Director of the United Nations Office for Outer Space Affairs (UNOOSA) from 2014 to 2022, I saw global space activities accelerate at an impressive pace, and I also saw the wide and growing need for a broad understanding of the global governance of space, which is why I'm pleased to see this latest iteration of SWF's *Handbook for New Actors* in Space. Chapter 1 should serve as a solid foundation and a good jumping-off point for more specialized understandings of space governance, including what is explained in later chapters, and in the numerous references given throughout this latest version of the Handbook.

CHAPTER ONE
The International Framework
for Space Activities

FREEDOM AND RESPONSIBILITY

Three core principles lie at the heart of the international framework for space activities: freedom of exploration and use of space, peaceful purposes, and state responsibility. These principles, as contained in the five core treaties, form the foundation of international space law and are reflected in many of the other legal and political mechanisms that make up the international framework for space activities. The following sections provide an overview of each principle.

Freedom of Exploration and Use of Space

Outer space is free to be explored, and no nation or state can restrict another state's legitimate access to space for peaceful purposes. This freedom is enshrined in the most important source of space law, the Treaty on Principles Governing the Activities of States in the Exploration and Use of Outer Space, including the Moon and Other Celestial Bodies, more commonly referred to as the Outer Space Treaty.

Figure 1.1 | A close look at one of the original copies of the Outer Space Treaty.
Credit: Max Alexander

Like all treaties, the Outer Space Treaty balances rights with obligations. The freedoms to use and explore space are balanced with the obligations listed throughout the treaty. Those obligations can be considered positive obligations requiring a state to perform certain actions, or negative obligations that prohibit actions. Article I of the Outer Space Treaty lists these all-important freedoms, explaining that:

> Outer space, including the Moon and other celestial bodies, shall be free for exploration and use by all States without discrimination of any kind, on a basis of equality and in accordance with international law, and there shall be free access to all areas of celestial bodies.

This free access means that new and emerging actors in space have just as much right to explore and use space for peaceful purposes as the established space actors. The preceding clause of Article I also directly states that the activity of exploring and using outer space is the "province of all mankind."

The Outer Space Treaty then requires that "[t]here shall be freedom of scientific investigation in outer space, including the Moon and other celestial bodies, and States shall facilitate and encourage international cooperation in such investigation." Indeed, the very nature of the Outer Space Treaty encourages international cooperation and scientific investigations as ways to promote peace and stability among the nations of the world.

Like most international treaties, the preamble to the Outer Space Treaty does not contain legally operative language establishing rights, obligations, or prohibitions. Rather, it contains the object and purpose of the treaty—the subject matter being addressed, the reason the treaty is being drafted, and what the treaty is intended to establish. The preamble to the Outer Space Treaty explains the motives and aspirations behind the creation of the treaty, formalizing the reasons that states decided to create it; these being because they:

- recognize the common interest of all humankind in the progress of the exploration and use of outer space for peaceful purposes;

- believe that the exploration and use of outer space should be carried out for the benefit of all peoples, irrespective of their degree of economic or scientific development;

- desire to contribute to broad international development of both the scientific and legal aspects of space exploration and use; and

- believe that this international cooperation will drive mutual understanding and strengthen friendly relations among states and peoples.

These beliefs in the preamble to the Outer Space Treaty reflect the intentions of the drafters who created this international legal instrument. All international space law should be read with the understanding that these are the intentions and aspirations behind the Outer Space Treaty. No interpretation of space law (whether that law is international or national) should circumvent, subvert, or defeat the motives and purposes listed above. In fact, any valid interpretation of any of the articles of the Outer Space Treaty must reflect, conform, and serve these purposes. These aspirations, contained in the preamble but forming an integral part of the treaty, should always be remembered when considering the freedom to access space, explore space, or partake in any other activity or use of space.

> The freedom to explore outer space is held by all states, and through them, by all peoples of the world. No state can lawfully prevent or restrict any new entrant to the field of peaceful space activities.

Additionally, it should be noted that the words "exploration" and "use" are in the very title of the Outer Space Treaty. The use of outer space, including the use of the Moon and the use of any celestial bodies, was contemplated by the drafters and negotiators of the treaty, and is part of the freedom of access, exploration, and use as codified in Article I. Consequently, the freedom to explore outer space is held by all states, and through them, by all peoples of the world. No state can lawfully prevent or restrict any new entrant to the field of peaceful space activities.

While some treaties may address space activities in a tangential fashion, there are five core treaties, listed in Table 1.1, that address space activities specifically.

Treaty	Adoption by General Assembly	Entered into Force	Number of Ratifying States*
Treaty on Principles Governing the Activities of States in the Exploration and Use of Outer Space, including the Moon and Other Celestial Bodies (Outer Space Treaty)	1966	1967	115
Agreement on the Rescue of Astronauts, the Return of Astronauts and the Return of Objects Launched into Outer Space (Astronaut Agreement)	1967	1968	100
Convention on International Liability for Damage Caused by Space Objects (Liability Convention)	1971	1972	100
Convention on Registration of Objects Launched into Outer Space (Registration Convention)	1974	1976	75
Agreement Governing the Activities of States on the Moon and Other Celestial Bodies (Moon Agreement)	1979	1984	16

*As of May 2024

Table 1.1 | The core treaties on space

Core Treaties

The core space treaties were negotiated and drafted by the United Nations (UN) Committee on the Peaceful Uses of Outer Space (COPUOS), a standing body of Member States of the United Nations that has considered the political, legal, and scientific aspects of space activities since the beginning of the space age. The titles of the treaties in Table 1.1 illustrate their basic subject matter, and the four treaties adopted after the Outer Space Treaty largely elaborate upon and refine provisions of the foundational Outer Space Treaty. The 1968 Astronaut Rescue and Return Agreement refines and expands on the protection given to astronauts, while the 1972 Liability Convention similarly expands the provisions for liability for damage incurred in the launching and operation of space objects. The Liability Convention establishes absolute liability for physical damage suffered on the surface of the Earth, or to aircraft in flight, and establishes a fault-based liability regime for space objects in outer space. The 1975 Registration Convention makes mandatory both international registration and the establishment of national registries of space objects.

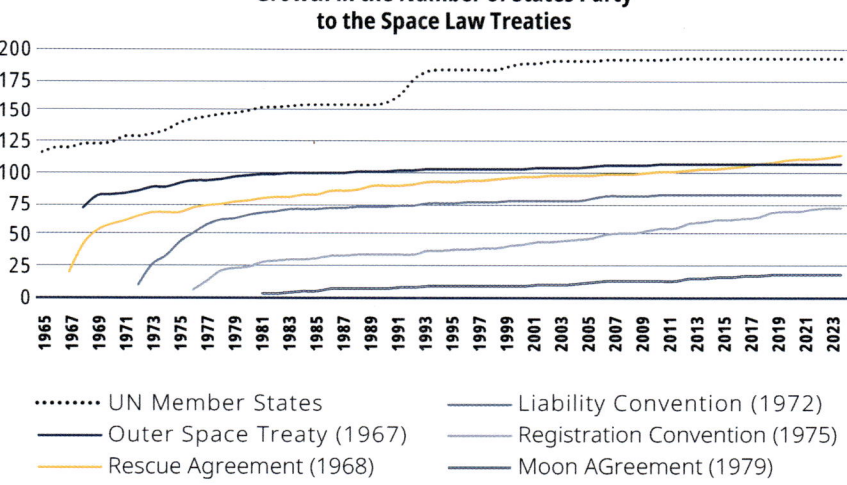

Growth in the Number of States Party to the Space Law Treaties

.... UN Member States
——— Outer Space Treaty (1967)
——— Rescue Agreement (1968)
——— Liability Convention (1972)
——— Registration Convention (1975)
——— Moon AGreement (1979)

Figure 1.2 | The figure above illustrates the number of states party to the space law treaties as of May 2024.

The core UN treaties on space were all drafted from the mid-1960s until the late 1970s. This era of broad treaty-making by the United Nations is now over, and subsequent decades have seen the United Nations use General Assembly use resolutions to communicate principles on a number of subsequent space-related topics, such as direct television broadcasting, remote sensing from space, the use of nuclear power sources in space, aspects of the use of geostationary orbit, and enhanced practices in the registration of space objects, and other topics.

While these documents are not strictly binding in the way that treaties are, they have significant normative value—especially as they were developed within COPUOS, adopted through consensus, and many of them were then subsequently endorsed at the UN General Assembly. They have served as indications of state positions on a variety of topics, and have served as the basis for national policy and regulation.

Year	Title	UN Doc
1961	International Cooperation in the Peaceful Uses of Outer Space	Res. 1721 A and B (XVI)
1963	Declaration of Legal Principles Governing the Activities of States in the Exploration and Use of Outer Space	Res. 1962 (XVII)
1982	Principles Governing the Use by States of Artificial Earth Satellites for International Direct Television Broadcasting	Res. 37/92
1986	Principles Relating to Remote Sensing of the Earth from Outer Space	Res. 41/65
1992	Principles Relevant to the Use of Nuclear Power Sources in Outer Space	Res. 47/68
1996	Declaration on International Cooperation in the Exploration and Use of Outer Space for the Benefit and in the Interest of All States, Taking into Particular Account the Needs of Developing Countries	Res. 51/122
2000	International Cooperation in the Peaceful Uses of Outer Space	Res. 55/122
2004	Application of the Concept of "Launching State"	Res. 59/115
2007	Recommendations on Enhancing the Practice of States and International Intergovernmental Organizations in Registering Space Objects	Res. 62/101
2007	Space Debris Mitigation Guidelines of the Committee on the Peaceful Uses of Outer Space	Res. 62/217
2009	Safety Framework for Nuclear Power Source Applications in Outer Space	
2013	Recommendations on National Legislation Relevant to the Peaceful Exploration and Use of Outer Space	Res. 68/74
2019	Guidelines for the Long-term Sustainability of Outer Space Activities of the Committee on the Peaceful Uses of Outer Space	A/74/20

Table 1.2 | Important UN declarations on space

Article III of the Outer Space Treaty incorporates space law into the larger body of international law. Consequently, other sources of public international law, including the UN Charter, impact the law of outer space. The practices of states, along with general principles of law, are also valid and often applicable. For example, one principle of general international law can be summarized as "that which is not explicitly prohibited is otherwise permitted." The consequence of these explicit freedoms, and their context in the larger body of international law, is the creation of a wide scope of state freedom in outer space with only certain particular and explicitly codified legal prohibitions.

For states looking to begin their first forays into conducting space activities, signing and ratifying the core treaties sends a signal to the world that the rights and obligations of international space law are understood and accepted, and underlies their serious approach to beginning space activities. It shows that they intend to be a responsible and law-abiding actor in space, and that they have "joined the club" of spacefaring nations.

Peaceful Purposes

Article IV of the Outer Space Treaty requires that states refrain from placing nuclear weapons or other weapons of mass destruction into Earth orbit or installing or stationing them on celestial bodies. It further requires that the Moon and other celestial bodies be used exclusively for peaceful purposes. Next, it forbids the establishment of military bases, installations, or fortifications on celestial bodies, and also forbids testing weapons and conducting military maneuvers on celestial bodies. Another treaty, the Partial Test Ban Treaty of 1963, prohibited states from testing nuclear weapons or performing nuclear explosions beyond the limits of the atmosphere, including in outer space, while another— the Comprehensive Test Ban Treaty of 1996—prohibits any nuclear weapon test explosions.

For states looking to begin their first forays into conducting space activities, signing and ratifying the core treaties sends a signal to the world that the rights and obligations of international space law are understood and accepted, and underlies their serious approach to beginning space activities.

An important point is that this focus on peaceful purposes is not a prohibition on military uses of space. There have been military and security aspects to space activities since the very beginning of the space age. As a foundational security treaty negotiated between Cold War powers, the Outer Space Treaty addresses this dual-use nature of space capabilities. Since the treaty entered into force, there has always been a debate about the definition of peaceful purposes, with two main interpretations arising: one says that peaceful purposes means "nonmilitary" in any regard; the other holds that peaceful merely means "nonaggressive." The latter interpretation has gradually gained broader acceptance. However, the explicit prohibitions mentioned above remain, and there are still debates today over the interpretation of the peaceful purposes principle.

As other sources of international law are also applicable to space activities through their inclusion in Article III of the Outer Space Treaty, the general prohibition on the threat of, or use of, force between UN Member States is therefore applicable to outer space. Article 2.4 of the UN Charter requires that:

> [a]ll Members shall refrain in their international relations from the threat or use of force against the territorial integrity or political independence of any state, or in any other manner inconsistent with the Purposes of the United Nations.

Additionally, Articles 39 through 51 address threats and breaches of the peace, acts of aggression, and the inherent right of self-defense. This general regime of public international law between states underpins the special regime of space law and creates the same prohibitions and restrictions for military conflict in space as on Earth. There is also ongoing scholarly work on the specific applications of international law to conflict in space as exists in the maritime, air, and terrestrial domains, with projects such as the Manual on International Law Applicable to Military Uses of Outer Space (MILAMOS) and the Woomera Manual on the International Law of Military Space Activities and Operations.

International State Responsibility

In the usual affairs of humankind, governments are not generally responsible for the actions of their citizens. If a citizen of *state A* goes abroad to *state B*, and someone in *state B* wants to bring a claim against them, they don't often also name *state A's* government as a co-defendant. In the usual dealings between people and foreign governments, people are not the responsibility of their governments. This is not the case in outer space activities.

Under Article VI of the Outer Space Treaty, states are directly responsible for all their national space activities, whether that activity is conducted by the government itself or by any of its citizens or companies, and whether launching domestically or possibly even when its citizens are conducting space activities abroad.

Under Article VI of the Outer Space Treaty, states are directly responsible for all their national space activities, whether that activity is conducted by the government itself or by any of its citizens or companies, and whether launching domestically or possibly even when its citizens are conducting space activities abroad. The direct responsibility of national governments is relatively unique in international law. Article VI of the Outer Space Treaty begins with:

> *States Parties to the Treaty shall bear international responsibility for national activities in outer space, including the Moon and other celestial bodies, whether such activities are carried on by governmental agencies or by non-governmental entities, and for assuring that national activities are carried out in conformity with the provisions set forth in the present Treaty.*

The second sentence continues:

> *The activities of non-governmental entities in outer space, including the Moon and other celestial bodies, shall require authorization and continuing supervision by the appropriate State Party to the Treaty.*

Because the direct responsibility and potential international liability for all national activities is relatively unique and quite broad, this provision of the Treaty should always be taken into account by all space actors engaged in the authorization, supervision, planning, or conducting of space activities. The requirement that space activities be carried out in conformity with the treaty acts as a limiting provision to Article I's freedoms of access, exploration, and use. When space activities cause physical damage on the ground, to aircraft in flight, or to space objects in space, then mere international responsibility expands to international liability, a separate but related issue expanded upon in Chapter One: International Liability.

Today, many space activities are international in nature, and in any multinational space project, all states involved are under these obligations. This expansive international state responsibility is an incentive for developing and enacting national space policy and space legislation, the subject of Chapter Two.

UN COPUOS Guidelines for the Long-term Sustainability of Outer Space Activities (LTS Guidelines)

In 2019, following an eight-year process, the United Nations Committee on the Peaceful Uses of Outer Space (COPUOS) adopted a set of twenty-one guidelines reflecting global best practices in fostering and ensuring space sustainability. The Guidelines for the Long-term Sustainability of Outer Space Activities, referred to as the LTS Guidelines, are voluntary in nature and are non-legally binding. They are intended to support states developing space capabilities with the best existing practices for the safety of space operations and to avoid causing harm to the outer space environment.

The LTS Guidelines address policy, regulatory, operational, safety, scientific, technical, international cooperation, and capacity-building aspects of space activities. New actors in space are encouraged to familiarize themselves with the content of these guidelines as a means of ensuring that their space activities are conducted in a sustainable manner. Although non-binding, a growing number of states are adapting their national regulatory frameworks to implement aspects of these guidelines as a way of demonstrating their commitment to utilizing the space environment in a safe and sustainable manner.

REGISTRATION OF SPACE OBJECTS

Along with international responsibility for national space activities, and potential international liability for damage caused to other states, states are also obligated to register their space objects. The tracking of which states are responsible for which space activities is aided by the registration of space objects in both national and international registries of space objects.

International registration of space objects was first called for in United Nations General Assembly (UNGA) Resolution 1721 B (XVI), adopted by the UN at the dawn of the space age in 1961. This resolution calls upon states launching space objects to promptly provide the UN with launch information for a UN-maintained public registry. This international registry was intended to aid other states in determining which activities in space are being conducted by whom. Today this voluntary notification to the UN would be called a transparency and confidence-building measure (TCBM), as notifying the rest of the world about launches also helps show that a state is open about its activities.

While UNGA Res. 1721 B (XVI) is not legally binding and imposes no mandatory obligations on states, international registration of launched space objects was made mandatory in 1975 with the Registration Convention—at least for those states which are a party to that convention. As of May 2024, 75 states have ratified the Registration Convention. Although this is significantly fewer than the number of states that are party to the Outer Space Treaty, it includes all the major and historical space powers.

Articles III and IV of the Registration Convention require that the UN Secretary-General establish a registry of space objects with open access to all. Article IV requires that any launching state placing its launched space object on a national registry shall also communicate to the Secretary-General certain information for the international registry. That information is:

INFORMATION TO COMMUNICATE FOR REGISTRATION

- » The name of the launching state (or states)
- » An appropriate designator of the space object or its registration number
- » Date and territory or location of launch
- » Basic orbital parameters, including:
  - *Nodal period*
  - *Inclination*
  - *Apogee*
  - *Perigee*
- » General function of the space object

The remaining requirements include updating the UN with additional information and adding information on objects that are no longer in Earth orbit. On behalf of the Secretary-General, the United Nations Office for Outer Space Affairs (OOSA) is the keeper of this international registry established by the Registration Convention, as well as the registry of objects registered pursuant to UNGA Res. 1721 B (XVI). For states not party to the Registration Convention, international registration can be made pursuant to UNGA Res.1721 B (XVI). OOSA maintains a standard form for both registries, which it recommends that states use (see Figure 1.3). The required registration information is not overly detailed.

Part A:
Information provided in conformity with the Registration Convention or General Assembly Resolution 1721 B (XVI)

New registration of space object	*Yes* ☐	*Check Box*
Additional information for previously registered space object	*Submitted under the Convention: ST/SG/SER.E/* ☐ *Submitted under resolution 1721B: A/AC.105/INF.* ☐	*UN document number in which previous registration data was distributed to Member States*

Launching State/States/international intergovernmental organization

State of registry or international intergovernmental organization		*Under the Registration Convention, only one State of registry can exist for a space object.*
Other launching States		

Designator

Name		
COSPAR international designator		
National designator/registration number as used by State of registry		

Date and territory or location of launch

Date of launch (hours, minutes, seconds optional)	*hrs* *min* *dd/mm/yyyy* *sec*	*Coordinated Universal Time (UTC)*
Territory or location of launch		

Basic orbital parameters

Nodal period		*minutes*
Inclination		*degrees*
Apogee		*kilometres*
Perigee		*kilometres*

Figure 1.3 | These figures illustrate what information appears in the UNOOSA registration form. This representation was created by Secure World Foundation.

Part A:
Information provided in conformity with the Registration Convention or General Assembly Resolution 1721 B (XVI)

General function

General function of space object

Change of status

Date of decay/reentry/deorbit (hours, minutes, seconds optional)	*dd/mm/yyyy*	*hrs* *min* *sec*	*Coordinated Universal Time (UTC)*

Sources of information

UN registration documents	*http://www.unoosa.org/oosa/SORegister/docsstatidx.html*
COSPAR international designators	*http://nssdc.gsfc.nasa.gov/spacewarn/*
Global launch locations	*http://www.unoosa.org/oosa/SORegister/resources.html*
Online Index of Objects Launched into Outer Space	*http://www.unoosa.org/oosa/osoindex.html*

Part B:
Additional information for use in the United Nations Register of Objects Launched into Outer Space, as recommended in General Assembly Resolution 62/101

Change of status in operations

Date when space object is no longer functional (hours, minutes, seconds optional)	*dd/mm/yyyy*	*hrs* *min* *sec*	*Coordinated Universal Time (UTC)*
Date when space object is moved to a disposal orbit (hours, minutes, seconds optional)	*dd/mm/yyyy*	*hrs* *min* *sec*	*Coordinated Universal Time (UTC)*
Physical conditions when space object is moved to a disposal orbit (see COPUOS Space Debris Mitigation Guidelines)			

Basic orbital parameters

Geostationary position (where applicable, planned/actual)		*degrees East*

Additional Information

Website:	

Part C:
Information relating to the change of supervision of a space object, as recommended in General Assembly Resolution 62/101

Change of supervision of the space object

Date of change in supervision *(hours, minutes, seconds optional)*	dd/mm/yyyy	hrs / min / sec	*Coordinated Universal Time (UTC)*
Identity of the new owner or operator			

Change of orbital position

Previous orbital position		*degrees East*
New orbital position		*degrees East*
Change of function of the space object		

Part D:
Additional voluntary information for use in the United Nations Register of Objects Launched into Outer Space

Basic information

Space object owner or operator	
Launch vehicle	
Celestial body space object is orbiting (if not Earth, please specify)	
Other information (information that the State of registry may wish to furnish to the United Nations)	

Sources of information

General Assembly resolution 62/101	*http://www.unoosa.org/oosa/SORegister/resources.html*
COPUOS Space Debris Mitigation Guidelines	*http://www.unoosa.org/oosa/SORegister/resources.html*
Texts of the Registration Convention and relevant resolutions	*http://www.unoosa.org/oosa/SORegister/resources.html*

UNGA Resolution 62/101

The registry form (Figure 1.3) also references UNGA Res. 62/101 from 2007 entitled, "Recommendations on enhancing the practice of States and international intergovernmental organizations in registering space objects." The resolution expresses a desire for states to proffer additional information regarding space objects, including updated circumstances such as a change of function, change of operational status, change of orbital position, removal to a disposal orbit, or change in the owner and/or operator of the registered space object. This ability to update information provided to the UN is a key advancement and has implications for more advanced or complex space activities such as launches with multiple launching states, and for satellite servicing, satellite replacement in orbital constellations, or debris removal in the future.

Online UN Index of Objects Launched Into Outer Space

UNOOSA maintains an online index of objects launched into outer space. The online index is a searchable database and can be a useful first source of information about a nation's launched and registered space objects, with inclusion of the details and particulars in the form listed above.

The Outer Space Objects Index allows users to search using a variety of parameters, including: "State or Organization" (which has launched or procured the launch of the object); "Launch Facility" (i.e., location of the launch); and "Year" (of launch). The online index also employs a number of other interesting and useful filters, such as "Status," which reflects the approximate location or destination of the space mission with categories such as "in orbit" for an object orbiting Earth, "decayed," "deorbited," "on Mars," and even "interstellar." Another filter is "In Orbit," where the user can either select "Yes" or "No," which is a useful way to refine one's search. Another useful filter is "UN Registered," reflecting whether the space object has been officially registered with the UN (or whether UNOOSA has simply entered the information, absent direct notification).

As a general transparency measure, the UNOOSA online index of space objects can be a useful first stop when answering questions about which state or states are internationally responsible for which space object.

National Registration

Article VIII of the Outer Space Treaty does not address international registration. Rather, it discusses national registration, stating that

> [a] State Party to the Treaty on whose registry an object launched into outer space is carried shall retain jurisdiction and control over such object, and over any personnel thereof, while in outer space or on a celestial body. Ownership of objects launched into outer space, including objects landed or constructed on a celestial body, and of their component parts, is not affected by their presence in outer space or on a celestial body or by their return to the Earth.

In an area where state sovereignty is absent, the effect of this article is to provide a crucial component of state sovereignty, namely jurisdiction. The right of a state to exercise jurisdiction over space objects depends upon that state listing its launched objects on a national registry. Each state might need to consolidate that international right in its national legislation.

Enshrining in an international treaty the national right to exercise jurisdictional powers in an extraterritorial manner through a national registry gives states an incentive to establish national registers, and to list their space objects on them. In doing so, it furthers the transparency of space activities, and as long as national registries are publicly searchable, outsiders can determine which space objects belong to which country. Coupled with this are the final sections of Article VIII, whereby states retain ownership of their launched space objects and their component parts while in outer space and upon return to Earth. States party to the Outer Space Treaty should consider establishing and maintaining a national space registry because the Registration Convention makes the creation of such national registries obligatory.

As of May 2024, 39 states have national space registries, and some make their national registries available and searchable online. While international organizations cannot be parties to the Registration Convention, the European Space Agency (ESA) and the European Organisation for the Exploitation of Meteorological Satellites (EUMETSAT) also keep registries of their space objects. As the method for exercising jurisdiction over launched space objects, the national registry is an important component of a state's oversight and responsibility requirements. National registration is discussed further in Chapter Two: National Registration.

Suborbital Launches

The Registration Convention requires registration of objects "launched into Earth orbit or beyond," and the previous UNGA resolution likewise calls upon states launching objects into orbit or beyond to furnish information through the Secretary-General for the registration of launchings. However, there is no international requirement or call to register objects that are only being launched for suborbital operations. How to deal with suborbital space activities is an open question that new actors will need to consider from a registration perspective, as registration may impact whether suborbital activities are considered to be their national space activities, incurring international responsibility. To date, many states have not made a legal determination whether and to what extent international space law is applicable to suborbital activities.

However, one of the main goals of international registration is to alert the world to a state's space activities. Consequently, continuing to observe the above mentioned international registration requirements fulfills these objectives of international transparency and confidence-building about national space activities.

INTERNATIONAL FREQUENCY MANAGEMENT

Spacecraft communicate using the electromagnetic spectrum, and nearly all of them use the radio-frequency part of the spectrum that is shared with all other space and terrestrial users. Consequently, frequency coordination and allocation among users is one of the most important processes for the successful operation of a space project.

The International Telecommunication Union (ITU) is a specialized agency of the United Nations. The oldest organization within the UN system, the ITU traces its origin to international telegraph unions in the mid-19th century. Today, the ITU has 193 member states that are party to its principal treaties: the ITU Constitution and the ITU Convention. Since the beginning of the space age, the ITU has aided the exploration and use of space through international coordination and frequency allocation. The ITU is tasked with ensuring the rational, equitable, efficient, and economical use of the radio-frequency spectrum. Within the ITU, this task is primarily managed by the ITU Radiocommunication (ITU-R) sector. The ITU-R also keeps track of which orbital locations are using those frequencies, either in specific "slots" along the Equatorial axis assigned to satellites in geostationary Earth orbit (GEO), or in altitudes and planes used by non-geostationary satellites.

GEO is a limited natural resource defined as a circular orbit 35,786 kilometers above the Equator where satellites appear to hover over a fixed position on the surface of the Earth. Non-geostationary satellites, by default, occupy all the other altitudes, most usually those in medium or low Earth orbits (between 500 and 15,000 kilometers). Any satellites using the same frequency in proximity to other users (satellites or terrestrial) can cause interference to each other, thus the ITU Radio Regulations require coordination between users to prevent harmful interference.

The ITU-R maintains the ITU Radio Regulations (the "Radio Regs"), which are updated roughly every four years when the Member States of the ITU gather for the World Radiocommunication Conference (WRC) whose output is considered a treaty-level agreement. The ITU also maintains the Master International Frequency Register (MIFR) of all coordinated frequencies. The MIFR, along with the Frequency Allocation Table maintained by the pertinent nation, should be consulted very early on in a space project, when considering which frequency or frequencies a space project's space systems and Earth stations will use.

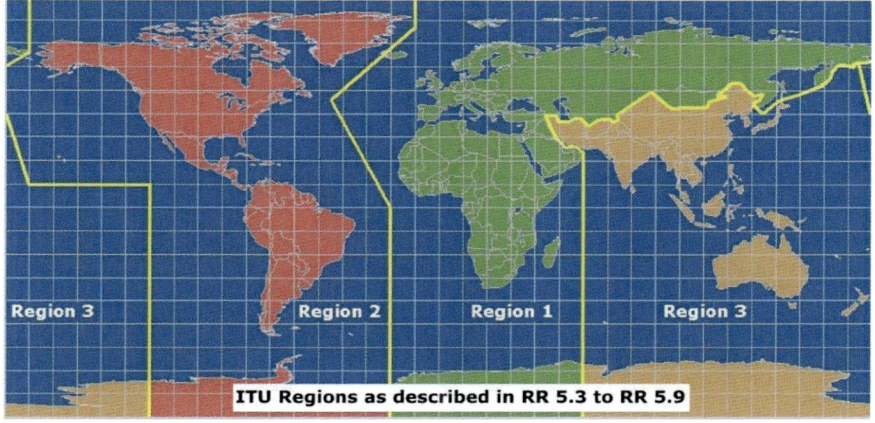

Figure 1.4 | This figure illustrates the three regions for radio regulations, as defined by the ITU. *Credit: ITU*

The ITU-R divides the world into three regions for spectrum allocations, as shown in Figure 4. Region 1 includes Europe, Africa, the former countries of the Soviet Union, and Mongolia. Region 2 includes the Americas and Greenland. Region 3 is the rest of Asia, Australasia, and the Pacific. Each spectrum region has allocated particular frequencies to particular technologies and services for terrestrial and space applications. The ITU has allocated a number of frequencies for specific space activities, including frequencies for Earth exploration, meteorology, radio astronomy, emergency telecommunications, radio navigation, space operations, space research, and amateur satellites.

The radio-frequency spectrum is divided into bands that are either exclusively allocated to specific types of applications or that share allocations for various applications. Applications with broad international usage enjoy harmonized allocations in all three regions. A shared portion of the spectrum is available for one or more services, either on a worldwide or regional basis. Within the shared bands, different services are classed into either primary or secondary services. Primary services enjoy superior rights to secondary services.

The Radio Regulations require that secondary services:

- not cause harmful interference to stations of primary services to which frequencies are already assigned or may be assigned at a later date;

- cannot claim protection from harmful interference from stations or a primary service to which frequencies are already assigned or may be assigned at a later date; and

- can, however, claim protection from harmful interference from stations of the same or other secondary services to which frequencies may be assigned at a later date.

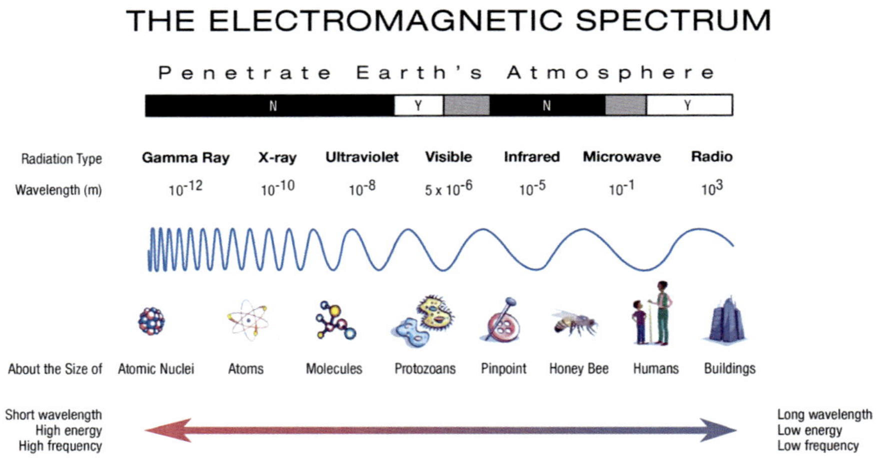

THE ELECTROMAGNETIC SPECTRUM

Penetrate Earth's Atmosphere

	N		Y		N		Y

Radiation Type	Gamma Ray	X-ray	Ultraviolet	Visible	Infrared	Microwave	Radio
Wavelength (m)	10^{-12}	10^{-10}	10^{-8}	5×10^{-6}	10^{-5}	10^{-1}	10^{3}

About the Size of	Atomic Nuclei	Atoms	Molecules	Protozoans	Pinpoint	Honey Bee	Humans	Buildings

Short wavelength
High energy
High frequency

Long wavelength
Low energy
Low frequency

Figure 1.5 | The chart illustrates the electromagnetic spectrum. *Credit: NASA*

Figure 1.5 shows the distribution of applications into different parts of the spectrum, each of which also depends on the region of the world in which that use is located. National administrators implement and apply the ITU Radio Regulations on the national level. A deeper discussion follows in Chapter Two, dealing with the procedure of coordinating with the ITU through national administrators, and in Chapter Three, dealing with coordination between operators and national administrators, and among the operators themselves.

World Radiocommunication Conference

World Radiocommunication Conferences (WRC) are held every three to four years, under the auspices of the ITU-R. Their purpose is to allow member states to review and revise the treaty deciding use of the radio-frequency spectrum and of the geostationary satellite and non-geostationary satellite orbit. A month-long conference with thousands of participants, the WRC is the primary venue through which frequency assignments for terrestrial, aerial, and space-based applications are reviewed and made. As such, decisions taken at the WRC can have significant impact on the spectrum resources available to satellite operators. In fact, the Radio Regulations are updated after each WRC to contain the complete texts of the Radio Regulations adopted by the World Radiocommunication Conference of 1995 (WRC-95) as revised and adopted by subsequent World Radiocommunication Conferences, including all Appendices, Resolutions, Recommendations, and ITU-R Recommendations incorporated by reference.

The WRC also determines the "Questions" for examination by the Radiocommunications Assembly and its Study Groups in preparation for future WRCs. Because agendas and questions are set years in advance, space actors should determine what areas being studied might affect their project plans and spectrum needs and whether they themselves need to advocate for changes to the Radio Regulations to accommodate their future plans.

Companies and other interested parties can become sector members of the ITU, allowing them to observe meetings and provide industry perspectives to the Study Groups and Working Parties.

In addition to the coordination efforts carried out under the aegis of the ITU, an additional notable group is the Space Frequency Coordination Group (SFCG), which is an informal group of frequency managers from civil space agencies established in 1980 to facilitate the exchange of information among the agencies on future plans for space systems or for other systems potentially affecting space systems. More details on the SFCG are contained in the Annex.

Laser Communications

In recent years, there has been considerable advancement in the development of laser communications systems for satellites. Unlike radio communications, which utilize signals in the radio part of the electromagnetic spectrum, laser communications utilize signals in the optical part of the electromagnetic spectrum. Laser technologies have been demonstrated for communicating between ground stations and satellites orbiting the Earth, between two satellites orbiting the Earth, and between satellites orbiting the Moon and Mars and ground stations on Earth.

There are several major differences between traditional radio satellite communications and laser communications. Most significantly, laser communications allow data transfer rates 10 to 100 times faster than radio frequency systems. Laser communications are line-of-sight, meaning that there must be a clear, direct line path between the transmitter and receiver.

This means that laser communications are not able to broadcast over a wide reception footprint. But this also makes laser communications much harder to intercept, and there is very little chance of unintentional interference. Laser communications can also support much higher bandwidth and data rates than radio waves.

Laser communications pose significant questions for international regulation. Under the current definitions adopted by the ITU, satellites utilizing laser communications are not restricted to a particular allocation of frequencies, and their member state might not require a license. The ITU member states currently only develop standards and regulations pertaining to radio-frequency spectrum, which does not include optical laser spectrum. However, there are some who feel the definition of "space-based communications" should be expanded to cover laser communications, as the assignment of spectrum licenses is currently one of the few ways to regulate space activities.

REMOTE SENSING

Each state enjoys sovereignty over its territory, and therefore states are often concerned about others gaining insight into what is happening within their territory, either for commercial, political, or military purposes. So while space is free to be explored and used, including freedom of overflight for spacecraft, many states feel some uneasiness about spacecraft turning their cameras back towards Earth, enabling neighbors to gain information.

To date, no international treaty directly governs remote sensing. Rather, a number of UNGA resolutions establish certain principles relevant to remote sensing. UNGA Resolution 41/65 of 1986 relates fifteen principles for states to follow in their remote sensing activities. The resolution first establishes a difference between "primary data" and "processed data." Primary data means those "raw data that are acquired by remote sensors borne by a space object and that are transmitted or delivered to the ground from space." Conversely, processed data means the "products resulting from the processing of the primary data." Analyzed information is defined as "information resulting from the interpretation of processed data, inputs of data, and knowledge from other sources."

Principle XII of UNGA 41/65 is perhaps the most important of the remote sensing principles, and strikes a balance between the freedom to explore space and the concerns states have about being observed ("sensed states"):

> As soon as the primary data and the processed data concerning the territory under its jurisdiction is produced, the sensed State shall have access to them on a non-discriminatory basis and on reasonable cost terms. The sensed State shall also have access to the available analysed information concerning the territory under its jurisdiction in the possession of any State participating in remote sensing activities on the same basis and terms, taking particularly into account the needs and interests of the developing countries.

While Resolution 41/65 is a non-binding resolution from the United Nations General Assembly, it is meant to reflect the best practices of spacefaring states. Beyond this resolution, data sharing has become a key principle in remote sensing activities because of an early recognition of the links between accessibility to such data and societal benefits, scientific progress, and commercial applications.

Open data exchange at the international level has been upheld especially for global meteorological data and related products, as adopted in World Meteorological Organization (WMO) Resolution 40 and is promoted actively by the Group on Earth Observations and the Committee on Earth Observation Satellites. For more information on these groups, see Annex. Additionally, the International Charter Space and Major Disasters is a global collaboration among seventeen space agencies and other governmental organizations, through which satellite-derived information and products are made available to support disaster response efforts.

INTERNATIONAL STANDARDS

International standards are used in many fields in order to increase safety, reliability, and quality, and are increasingly being implemented in the space field. A standard is a document that provides requirements, specifications, guidelines, or characteristics that can be used consistently to ensure that materials, products, processes, and services are fit for their purposes. Standards can be as specific as an outline on how to interface with a particular class of device, or as general as providing details around management best practices for ensuring quality.

While standards can be developed by any organization or entity, international standards are becoming increasingly important in a more globalized world. Adopting an international standard can help ensure compatibility across entire global sectors and can also be used by companies to signal to potential customers that their products or services are high-quality. Multiple organizations in a sector can use standards to codify lessons learned from past mistakes to help improve overall safety in a sector.

The development of standards for the space sector has often been led by international non-governmental and industry-organized organizations. These include the International Organization for Standardization (ISO), the Consultative Committee for Space Data Systems (CCSDS), the International Committee on Global Navigation Satellite Systems (ICG), the United Nations Committee of Experts on Global Geospatial Information Management (UN-GGIM), and the Telecommunication Standardization Sector (ITU-T). For more information on these organizations and how they develop standards for space, see Annex. The United States Office of Space Commerce also provides the Space Industry Technical Standards Compendium for reference.

INTERNATIONAL EXPORT CONTROL

There is significant international concern over the uncontrolled spread of both conventional military goods and technologies and dual-use technology such as space technology. Dual-use technology is commonly defined as technology having both civil and military applications. An example in the space industry is the chemical rocket, which can be used as a space launch vehicle to place satellites and humans into orbit, but which can also serve as a ballistic missile for delivering weapons of mass destruction. All new actors, including private non-governmental space actors, should be acutely aware of the sensitive nature and politically charged context of all space activities. There are several international export control measures in place to regulate the proliferation of ballistic missiles and their supporting technologies as well as the transfer of other sensitive dual-use technologies. These international measures are described below. Chapter Two provides examples of similar measures employed at the national level—including the U.S. export control regime, International Traffic in Arms Regulations (ITAR).

Wassenaar Arrangement

At the international level, the Wassenaar Arrangement on Export Controls for Conventional Arms and Dual-Use Goods and Technologies is a significant effort to control the proliferation of specific types of military and dual-use goods and technologies. It was established in 1996 and as of May 2024 has 42 participating states, mostly located in North America and Europe. The goal of the Wassenaar Arrangement is to contribute to regional and international security and stability by promoting transparency and greater responsibility for transfers of conventional arms and dual-use goods and technologies, thus preventing destabilizing accumulations. Participating states control items in the "List of Dual-Use Goods and Technologies and Munitions List" and work to prevent unauthorized transfers of those items. The arrangement also uses export controls as a way to combat terrorism, and is not designed to work against any particular state or group of states. Participating states agree to exchange information on sensitive dual-use goods and technologies, follow agreed-upon best practices, and report any transfers or denied transfers of controlled items made to recipients outside of the Wassenaar Arrangement.

Missile Technology Control Regime

The Missile Technology Control Regime (MTCR) is another important international control in the realm of space activities. The MTCR is a voluntary regime that was originally established in 1987, and in May 2024 had 35 participating countries. Four additional countries have agreed to abide by MTCR export control rules but have not formally joined. The goal of the MTCR is to coordinate national export licensing efforts in order to prevent the proliferation of delivery systems capable of delivering weapons of mass destruction.

Hague Code of Conduct

In 2002, the Hague Code of Conduct against Ballistic Missile Proliferation, also known as the Hague Code of Conduct, was created to augment the MTCR. The Hague Code of Conduct calls on participating states to exercise restraint in the testing, production, and export of ballistic missiles. While the Hague Code is less restrictive than the MTCR, with 145 participating states as of May 2024 it has significantly more international acceptance, and it serves as a solid transparency and confidence-building measure (TCBM). Subscribing states agree to making pre-launch notifications and annual declarations of their policies.

INTERNATIONAL LIABILITY

In international law, liability is a concept related to, but distinct from, responsibility. Article VII of the Outer Space Treaty establishes the obligation that states launching space objects shall be internationally liable for

> damage to another State Party to the Treaty or to its natural or juridical persons by such object or its component parts on the Earth, in air space or in outer space, including the Moon or other celestial bodies.

This obligation to be held liable for resulting damage is necessarily linked with responsibility, but is distinct enough to require close attention. Whereas responsibility, discussed above, is an obligation to ensure that all national activities are carried out in conformity with the Outer Space Treaty, the liability provision requires that states undertake action towards the compensation of other states should damage occur. The definition of damage, as contained in the 1972 Liability Convention, is

> loss of life, personal injury or other impairment of health; or loss of or damage to property or of persons, natural or juridical, or property of international intergovernmental organizations.

This definition is usually interpreted to mean actual physical damage, rather than pecuniary interests or other forms of non-physical damage.

Responsibility is placed on the state or states responsible for national activities. Liability may be imposed upon any "launching state" of space objects causing damage. While space launches are inherently dangerous, and the execution of a launch is not illegal per se, the imposition of liability for damage means that states shall offer compensation after damage occurs, with an understanding that no violation of international law is necessarily found if damage occurs.

The Outer Space Treaty defines four categories of launching state: (1) the state "that launches," (2) that which "procures the launching of a space object," and each state from whose (3) territory or (4) facility an object is launched. The Liability Convention and the Registration Convention reiterate these categories. Consequently, there may be more than one launching state for the purposes of liability. Indeed, this is how many space activities are conducted today.

> For states, the liability obligation means that while they are at liberty to conduct launches, they must ensure that they are otherwise lawful, and they must be ready to pay compensation to other states should certain damages occur (either on the ground, in the air, or in space).

For states, the liability obligation means that while they are at liberty to conduct launches, they must ensure that they are otherwise lawful, and they must be ready to pay compensation to other states should certain damages occur (either on the ground, in the air, or in space). While a launch may take place from another country's territory, a state may still be exposed to potential liability if its activities fall within one of the four broad categories of launching state. In multilateral space activities, it makes sense for state partners to contractually determine beforehand which of them they shall mutually consider as a state of registry.

UNGA Resolution 59/115 of 2004, *Application of the concept of "launching State,"* further addressed the often complex, multi-state aspect of modern launch activities, and recommended that states conducting space activities consider "enacting and implementing national laws authorizing and providing for continuing supervision of the activities in outer space of non-governmental entities under their jurisdiction." It also recommended that states "consider the conclusion of agreements in accordance with the Liability Convention with respect to joint launches or cooperation programmes;" as well as inviting COPUOS member States to "submit information on a voluntary basis on their current practices regarding on-orbit transfer or ownership of space objects."

Figure 1.6 | The Chandrayaan-3 spacecraft lifts off onboard a LVM-3 rocket from Satish Dhawan Space Centre. *Credit: ISRO*

Once a state is a "launching state" of a particular space object, it will always be considered a launching state of that space object And while there can be more than one *launching* state, there should usually be only one state that is the *state of registry*. It might seem that a launching state would always be the state of registry, but complex, multi-state international launches happen more and more frequently.

While being deemed a launching state is tied to the concept of liability, the state of registry is tied to oversight, licensing, authorization, supervision, and jurisdictional competency over the space object.

In short, states are both *responsible* for all their national space activities and potentially *liable* for activities in which they are considered the launching state. For new entrants to the field of space activities, these obligations mean that supervising states should seek to limit risky launches or those that might cause damage to other states. The supervising state may also put in place provisions to reduce or offset their potential exposure to liability, such as requiring that new non-governmental entrants find insurance for their proposed space activities should damage occur. Insurance is discussed in both Chapter Two and Chapter Three

DISPUTE SETTLEMENT

Though the desired outcome of any space activity would ideally never include a need for dispute resolution, either among states or private parties, or a combination of the two, it is essential to consider which dispute resolution mechanisms are available if needed. This section addresses the basic mechanisms of dispute resolution open to states and private parties.

Liability Convention

The 1972 Liability Convention provides a framework by which states can pursue claims for damage caused by a space object—to another space object, to aircraft in flight, or on the surface of Earth. It should be remembered that a liability claim is not an allegation that a launching state has violated international law, but merely that damage has occurred for which it is alleged to be responsible. The Liability Convention sets out specific parameters for diplomatic claim resolution, beginning in Article IX. According to Article X of the Liability Convention,

> *A claim for compensation for damage may be presented to a Launching State no later than one year following the date of the occurrence of the damage or the identification of the Launching State which is liable.*

Pursuit of a claim under the Liability Convention does not require the prior exhaustion of remedies in national courts. While a claim can be pursued either in national courts or through the Liability Convention, both avenues cannot be pursued concurrently.

If one or both parties to a dispute are not party to the Liability Convention, the Liability Convention does not apply. In that situation, any diplomatic resolution must follow the rules of international law that otherwise apply to the relevant states that are party to the dispute. For example, if both states are parties to the Outer Space Treaty, the provisions of Article VII of the treaty would apply.

Where a resolution cannot be achieved through diplomatic channels, the Liability Convention provides for the non-adversarial settlement of disputes in the context of a three-member claims commission, which can be initiated by any party to the dispute. The procedure for the formation of a claims commission is described in Articles XIV through XX. Whether they are resolved through diplomatic channels or through a claims commission, disputes decided under the Liability Convention are "determined in accordance with international law and the principles of justice and equity," which generally attempt to restore the state that suffered damage to the position they would have been in had the damage not occurred.

To date, there has never been an international case of damages brought under the Liability Convention that has led to the creation of a claims commission. In the late 1970s, the *Cosmos 954* incident in northern Canada resulted in Canada presenting a diplomatic claim for damages to the Soviet Union due to a nuclear-fueled satellite from the Soviet Union crashing over Canadian territory. However, this situation was settled without a clear precedent under the Liability Convention for a state suffering physical damage receiving compensation, as the Soviet Union paid Canada without accepting liability for the crash.

International Court of Justice

With regard to the settlement of space-related disputes between states, the International Court of Justice (ICJ) provides yet another option. Of course, the parties to a dispute must either agree to refer the dispute to the ICJ or recognize compulsory jurisdiction under the ICJ statute. Only states may bring claims to the ICJ. Meanwhile, Article 96 of the UN Charter stipulates that the UN General Assembly and the UN Security Council (and other UN organs and specialized agencies so authorized by the UNGA) may request advisory opinions from the ICJ.

While the ICJ has yet to decide a space-related case to date, it would arguably have subject-matter jurisdiction over any space dispute that would be considered a dispute of international law. A claimant state before the ICJ would allege that the respondent state has violated its international obligations—an assertion that is distinct from an allegation of mere liability for damage.

Arbitration and Mediation

Arbitration agreements usually take the form of a *compromis*, or a clause in a contract setting forth the rights and obligations of the parties. Such arbitration clauses are globally well recognized and are even favored in some jurisdictions, as they reduce the burden on court systems. However, not all parties share the same priorities for dispute resolution. An arbitration clause provides the parties with the authority to establish the arbitrator selection process and set arbitrator qualifications, and to determine whether and what discovery is available, what rules apply (evidentiary and procedural), scheduling, level of confidentiality, the role the arbitrators will serve, decision format and whether the decision is binding, the appeal process (if any), choice of law, provisional remedies, and methods of enforcement. Arbitration clauses can specify a particular arbitral tribunal, in which case the parties must comply with the rules and requirements of that tribunal.

Mediation, like both arbitration and adjudication, also employs neutral third parties to resolve a dispute. However, the mediator/s would not issue a binding decision. The procedures for mediation are less structured and more flexible than those followed by either courts or arbitral tribunals and can be entirely consensual or court-ordered. Resolution of disputes between non-governmental actors, such as corporations or other private entities, is dealt with in the following chapters.

In 2011, the Permanent Court of Arbitration (PCA), situated in The Hague, Netherlands, promulgated its Optional Rules for Arbitration of Disputes Relating to Outer Space Activities. Additionally, the PCA recommends a model clause for insertion into contracts. These rules establish an alternative means of settling disputes among states, international organizations, and private entities.

ENVIRONMENTAL ISSUES

In order to ensure their continued habitability and usability, it is necessary to protect both terrestrial and space environments. However, space activities such as launch are considered inherently dangerous and risky. Consequently, there are various international norms addressing environmental regulation which either forbid certain activities, or which establish international responsibility (including liability) for damage and environmental degradation.

Protection of the Earth Environment

Launching into space is an inherently dangerous activity, usually involving the combustion of large amounts of solid and/or liquid fuel and the rapid transit of advanced hardware through harsh and unforgiving environments. For that reason, launch sites are chosen in isolated places, far from where accidents can cause harm to others.

> In order to ensure their continued habitability and usability, it is necessary to protect both terrestrial and space environments.

A number of sources of law address protecting the Earth environment and allocate the burden of making compensation in case damage happens. On the international level, states are generally responsible for transboundary international harm they cause to other states. This obligation exists in the general custom of states, and is widely recognized. Particular to space law, Article VII of the Outer Space Treaty creates the liability rules for space launches, and includes liability for launching states causing damage to other states party to the treaty on the Earth, in air space, or in outer space.

Additionally, states are absolutely liable for damage their space launches cause on the surface of the Earth or damage to aircraft in flight. This absolute liability does not require that any fault or negligence be proven, merely that the damage occurred resulting from the activities of the responsible state. Consequently, while space activities are generally lawful, their ultra-hazardous nature is reflected in this absolute liability regime from the Outer Space Treaty and the Liability Convention.

Back Contamination of Earth

Article IX of the Outer Space Treaty largely concerns protecting the space environment, but the second sentence concerns protecting the Earth environment from space material. It reads:

> States Parties to the Treaty shall pursue studies in outer space, including the Moon and other celestial bodies, and conduct exploration of them so as to avoid their harmful contamination and also adverse changes in the environment of the Earth resulting from the introduction of extraterrestrial matter and, where necessary, shall adopt appropriate measures for this purpose.

The International Committee on Space Research (COSPAR) is an interdisciplinary science organization that has long been concerned with protecting the unique and pristine conditions of space environments—pristine, at least, in relation to humankind's interaction with them. To this end, COSPAR has promulgated planetary protection principles for space missions, and while the protection of other celestial bodies is discussed below, COSPAR's highest levels of precaution are recommended for Earth-return missions, which may cause so-called "back contamination."

COSPAR subdivides Earth-return missions into "Restricted Earth Returns" and "Unrestricted Earth Returns." The Unrestricted Earth Return classification applies to missions returning from celestial bodies such as the Moon and Venus, which have neither indigenous life forms nor the types of environments where life could flourish. Restricted Earth Return applies to missions returning from Mars and Europa, for example.

Use of Nuclear Power Sources in Space

Nuclear power sources have been used on spacecraft since the beginning of the space age. The steady and predictable decay of radioactive material gives off energy in amounts and in a manner suitable for a spacecraft's needs. Radioisotope thermoelectric generators (RTGs) and radioisotope heat units (RHUs) are historically proven methods of power generation, with both the United States and the Russian Federation utilizing such nuclear power sources in various space activities.

Recognizing the particular suitability of nuclear power sources for space missions, UNGA Resolution 47/68 of 1992 established 11 principles relevant to their use. The nuclear power principles reiterate the applicability of international law and the concepts and framework already established by the Outer Space Treaty and the Liability Convention regarding the responsibility for and potential liability of the launching state, and the jurisdiction and control of the registering state.

Principle 3 of the resolution discusses guidelines and criteria for use, stating that nuclear power sources in space shall be restricted to those missions that "cannot be operated by non-nuclear energy sources in a reasonable manner." It further requires that nuclear reactors shall only use highly enriched uranium-235 as fuel, and that reactors shall be designed and constructed so that they can only become critical upon reaching orbit or an interplanetary trajectory, and through no other way (including rocket explosion, re-entry, or impact with water or land).

Principle 5 contains instructions for making notifications about malfunctioning nuclear power sources that risk re-entry of radioactive materials to Earth. The information to be furnished includes basic launch and orbital parameters as well as information on the nuclear power source itself and the probable physical form, amount, and general radiological characteristics of the components likely to reach the ground. The notification should be sent to concerned states and to the UN Secretary-General. The principles further call for consultations and assistance between states, and reinforce the roles of responsibility, liability and compensation, and the settlement of disputes from the existing space treaties.

Subsequent to UNGA Res. 47/68, the Scientific and Technical Subcommittee of COPUOS worked jointly with the International Atomic Energy Agency to develop the Safety Framework for Nuclear Power Applications in Outer Space. This framework, though not legally binding, is intended to be used as a guide for national and intergovernmental safety purposes. The framework deals with the safe use of nuclear power sources in space missions, and contains guidance for governments on how to authorize space missions with nuclear power sources, guidance for the management of responsibility and safety roles of such missions, and technical guidance. When planned space missions involve nuclear power sources, these guidelines should be consulted early in the project.

Space Debris

After more than sixty-six years of space activities, humanity has created a significant amount of space debris. Space debris is generally defined as the non-operational satellites, spent rocket stages, and other bits and pieces created during the launch and operation of satellites. As of June 2024, there are currently more than 35,000 space debris objects being tracked in Earth's orbit by space situational awareness tracking systems. In addition, is estimated that, in total,

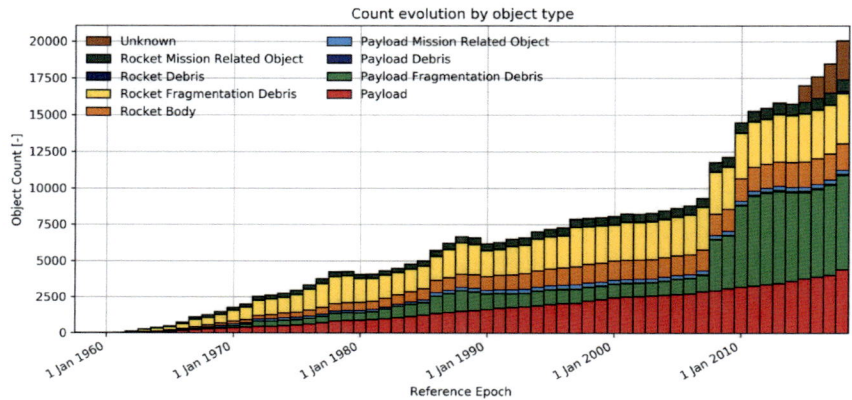

Figure 1.7 | The growth of space debris. *Credit: ESA*

there are 1,100,000 space debris objects of between 1 cm to 10 cm (0.4 to 4 inches) in size orbiting Earth. Each of these small, largely untracked objects could severely damage an active satellite in a collision. This debris is concentrated in the most heavily used regions of Earth orbit, where many active satellites also reside.

Former U.S. National Aeronautics and Space Administration (NASA) scientist Donald Kessler was one of the first to predict what has since become known as the Kessler Syndrome: as the amount of space debris in orbit grows, the increasing density of space debris will lead to random collisions between space debris. These random collisions would in turn generate more debris at a rate faster than it can be naturally removed from orbit by the Earth's atmosphere. Unlike the dramatic scenarios sometimes depicted in popular fiction, this process would take place much more slowly over decades or centuries.

There is now a general consensus among scientists that there is enough human-generated space debris concentrated in the critical region in LEO between 700 and 900 kilometers (430 to 560 miles) to create more debris even if no new satellites were launched. These debris-on-debris collisions will not lead to an infinite growth in the debris population. Rather, they will lead to a future equilibrium point that has a larger population of debris than today. The growth of debris will increase the risks—and thus the associated costs—of operating satellites in critical regions such as LEO. These increased costs could result from the need for more spare satellites to replace those lost in collisions, the need for heavier and more resilient satellites that cost more to build and launch, and increased operating costs resulting from trying to detect and avoid potential collisions.

Tracking of space debris, and efforts to reduce collision risk (and the creation of additional debris from collisions) is a key part of the purpose of space situational awareness systems and space traffic management and coordination efforts. Efforts to address space debris risk generally fall into two broad categories: mitigation efforts to reduce the creation of future debris through space operations; and remediation efforts to address the existing space debris population resulting from legacy activities.

Debris mitigation includes designing satellites and space systems to minimize the amount of debris they release during normal operations, developing methods to reduce the risk of fragmentation or explosion at end-of-life by venting leftover fuel or discharging batteries, and properly disposing of spacecraft and spent rocket stages after they are no longer useful. In recent years there has been increased focus on compliance with post-mission disposal timeline.

In the late 1990s, several major space agencies came together to form the Inter-Agency Space Debris Coordination Committee (IADC). The purpose of the IADC is to help coordinate and share research on space debris among participating space agencies. In 2007, the IADC published the first international Space Debris Mitigation Guidelines, which were updated most recently in June 2021 These technical guidelines define specific protected regions of Earth orbit and the recommended operational practices satellite operators should take to minimize

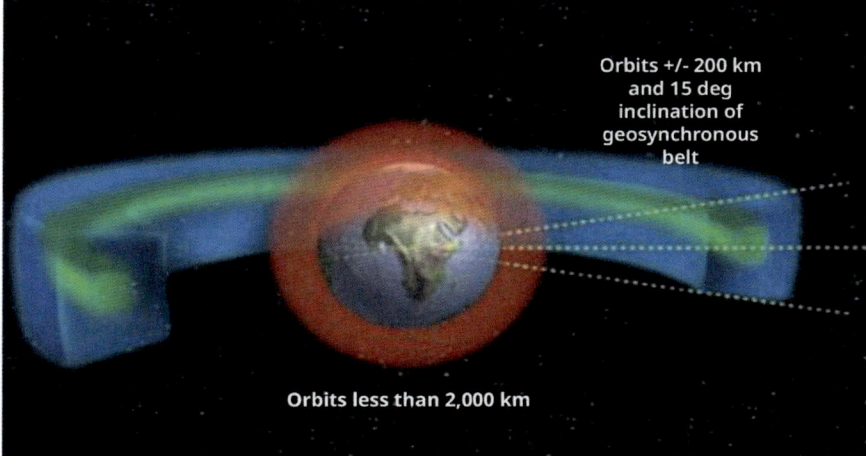

Orbits +/- 200 km and 15 deg inclination of geosynchronous belt

Orbits less than 2,000 km

Figure 1.8 | Depiction of protected regions as developed by the IADC. *Credit: ESA*

the creation of long-lived space debris in the protected regions. Figure 1.8 illustrates the various protected regions per the IADC guidelines.

A simplified subset of the IADC guidelines, the Space Debris Mitigation Guidelines of the Committee on the Peaceful Uses of Outer Space (which were more political in nature) were endorsed by the United Nations in 2009. Although the IADC guidelines remained voluntary, several states have implemented the debris mitigation guidelines through national regulations and policy, which is discussed in Chapter Two.

The IADC has also focused its research efforts on active debris removal (ADR). Several IADC member agencies are working on national ADR technology development efforts. In December 2022, the IADC released a statement on ADR efforts. This statement notes the importance of following through on post mission disposal guidelines to the greatest extent possible and suggests that new spacecraft and upper stages should be designed to ready for ADR, as a backup to other post-mission disposal plans. The statement also notes the need for further work on cost, risk, and benefit analysis of ADR technology.

Debris objects are of various sizes. Techniques and approaches to remediation of large debris will differ from techniques and approaches to remediation of small debris (1 to 10 centimeter size range). Technical experts from around the world have been working intensively on both of these problems over the last several years, and there are some promising technical solutions for removing either large objects or small objects. There is unlikely to be a single, all-encompassing solution that can deal with both large and small debris objectives.

Solving the challenges of space debris will require close coordination and cooperation among the engineers and scientists working on the technology, as well as the lawyers and policymakers developing policy and regulatory oversight. One step that can be taken is to avoid the intentional creation of space debris.

On April 19, 2022, U.S. Vice President Kamala Harris announced that the United States would voluntarily commit not to conduct destructive, direct-ascent anti-satellite (DA-ASAT) missile tests, and that the United States sought to establish this as a new international norm for responsible behavior in space. On December 12, 2022, the UNGA adopted Resolution 77/41, which called upon all States to commit not to conduct destructive DA-ASAT missile tests and to continue discussions in the relevant bodies to enhance space security. The resolution passed 155–9–9. As of June 2024, thirty-seven countries have made the same pledge not to conduct destructive DA-ASAT missile tests.

ADVANCED ISSUES

The preceding sections discussed important aspects of the international political, legal, and regulatory framework for space activities. Though subtleties exist at the boundaries of each of those topics, much is settled and understood. The last section of this chapter will discuss evolving issues and more advanced topics in space activities.

Boundary Between Airspace and Outer Space

Despite over half a century of space activities, there is no internationally recognized legal definition of where airspace ends and where outer space begins. Neither the Outer Space Treaty nor any other international legal instrument specifies a beginning or bottom point above which outer space begins. A definition of outer space is important because the legal regimes governing airspace and outer space are fundamentally different, and because getting to outer space requires crossing through airspace. A distinction between these two domains would help clarify which legal regime governs activities that cross between them.

Sovereignty, a fundamental component of the modern state, is essentially the power of a government to impose its exclusive authority—by creating laws, by deciding disputes, and through related powers such as enforcing its laws and judicial decisions. A state is exclusively sovereign in the airspace above its territory and territorial waters. However, Article II of the Outer Space Treaty severely undercuts state sovereignty in outer space, leaving only jurisdiction and ownership rights to a state's launched and registered space objects and personnel thereof. In air law, state sovereignty over airspace includes the right to keep others out, and only through complex bilateral and multilateral treaties do states allow civil aircraft from other states to enter (pass through, land on, and take off from) their sovereign airspace. This structure is the opposite of the regime for outer space; all states enjoy the right to freely access, explore, and use outer space.

In 1976, a number of countries in the equatorial regions of the globe signed on to the Bogotá Declaration, asserting a legal claim to control the use of space above their own territory. The declaration sought to upend the existing legal structure by

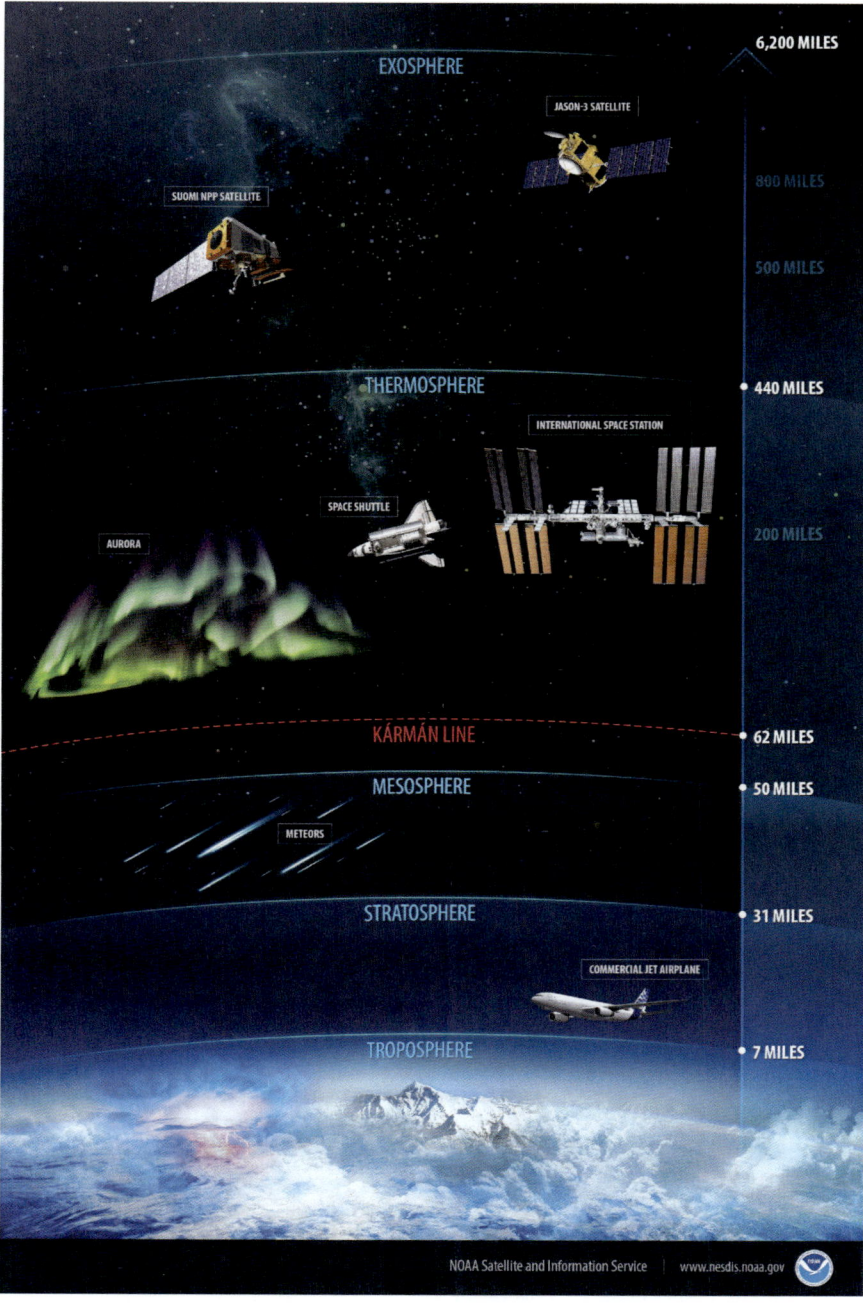

Figure 1.9 | Graphic depicting layers of Earth's atmosphere. *Credit: NOAA*

stating that the geostationary orbit, as a finite resource, "must not be considered part of the outer space." While Colombia's Constitution continues to recognize the orbital slot above the country as part of its territory, the Bogotá Declaration's claims have not been widely recognized and states continue to defer to ITU allocations of geostationary slots.

Some feel that this issue—the lack of a legal boundary between airspace and outer space—may increase in relevance in the near future. Some activities might be considered to be occurring in airspace and would therefore be governed by air law; alternatively, they might be considered to be occurring in outer space and thus would be governed by space law.

Is a reusable space plane governed by air law until it reaches Earth orbit? Or, because it is a spacecraft, does space law apply for the duration of the mission, including its transit through the atmosphere? As a general operational rule, it can be assumed that the area where artificial satellites are able to orbit Earth qualifies as outer space, although this altitude does not necessarily reflect the ceiling of airspace. A "spatialist" approach would argue for a bright-line distinction, perhaps at 100 kilometers above the Earth's surface—often called the Kármán line—where the atmosphere is too thin to sustain aeronautical flight.

Others first consider whether the activity involves craft with wings (like aircraft) or rockets (like spacecraft). Or they consider whether the craft takes off vertically (like a rocket) or horizontally (like a plane). Depending on whether the craft looks like an aircraft or spacecraft, and what its mission is, it might make sense to group it under air law or space law. This "functionalist" approach does not try to decide on a physical demarcation above the Earth's surface, as the spatialist approach recommends.

Whether something qualifies as an aviation activity or a space activity impacts, and is impacted by, not only the rules of the area where it operates, but which national rules it must inherently follow and which international responsibility and liability rules apply. To date, however, no international definition has been agreed upon. This lack of certainty might be a result of the previously clear distinction between aircraft in the air and rockets and satellites in outer space. Additionally, neither the functionalist nor the spatialist approach has dominated the discussion.

As technology develops and more states and non-state actors launch different types of craft and vehicles, it may become necessary to more clearly demarcate where space begins. A government considering space legislation might first consider whether there are any benefits to determining nationally where outer space legally begins, especially in the absence of an international definition. Likewise, non-governmental actors should be aware of the different regimes of air and space law, and the lack of international legal certainty between them.

Space Traffic Management

Space traffic management (STM) refers to measures taken to minimize or mitigate the negative impacts of the increasing physical congestion in space. As the number of active satellites and amount of space debris in space increases, particularly in highly used orbits and altitudes, physical congestion has become a growing problem. To date, there have been several confirmed, unintentional collisions between a functional satellite and another space object that have either damaged the satellite or completely destroyed both objects and created thousands of new pieces of space debris. The goal of STM is to try to eliminate future collisions and other incidents in space that could create additional debris or other safety risks for space activities, and to increase the safety and efficiency of space activities.

There is some debate over the term "Space Traffic Management," with space traffic coordination (STC) increasingly being used by some actors who feel that "management" suggests a higher level of active oversight or control than is possible, given that a very high percentage of space objects are uncontrolled space debris.

Space situational awareness (SSA) is an important element of STM. SSA refers to the ability to characterize the space environment and activities in space. A key component of SSA is using ground- or space-based sensors, such as radars or optical telescopes, to track space objects. The tracking data from multiple sensors is combined to estimate orbits for space objects and predictions of their trajectories in the future. Other key components include space weather, characterization of space objects, and pre-planned maneuvers as discussed in Chapter Three.

While some countries currently engage in practices that could be considered to be part of STM, there currently is no widespread state practice or established international regime. In 2010, the U.S. government began a program to provide close-approach warnings for all satellite operators. A few other countries provide similar warnings for national entities, and in 2023 the European Union's Space Surveillance and Tracking (EU SST) programme started providing a similar capability. Many commercial satellite operators work with a third-party service, such as the Space Data Association (SDA) or their own national space agency, to augment the basic warnings and data from governments. This is discussed more fully in Chapter Three.

There is ongoing debate over whether an international STM regime should begin with national practice or with an international treaty. Some have also made comparisons between STM and air traffic management and called for a new treaty to establish an international body that would set standards for STM and be similar to the function of the International Civil Aviation Organization (ICAO) for air traffic management. However, ICAO was created to resolve differences between previously existing national airspace regulations. Furthermore, the air traffic standards that ICAO sets require implementation by national regulative and administrative bodies, which many countries currently lack for space activities. As a result, others are pushing for major spacefaring states to establish national STM regimes that may evolve into an international regime in the future.

Dark and Quiet Skies

The last decade has seen a rapid increase and diversification of space activity, resulting in over ten thousand spacecraft operating in LEO as of May 2024. This notable proliferation of space objects has given rise to some worrying concerns, including interference with astronomical observations.

Figure 1.10, taken during a 333-second exposure from the Cerro Tololo Inter-American Observatory in Chile, shows a set of celestial images with Starlink satellites leaving nineteen individual

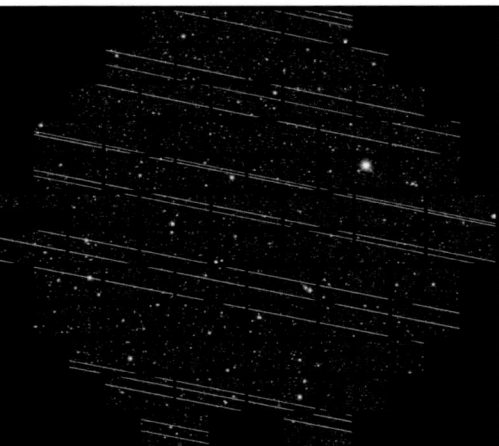

Figure 1.10 | Starlink satellites imaged from Cerro Tololo Inter-American Observatory (CTIO).
Credit: CTIO/NOIRLab/NSF/AURA/DECam DELVE Survey

streaks across the image. While computer programs can remove some satellite streaks, significant streaking renders many images useless. As observatories are oversubscribed for imaging time, ruined images like this mean lost time and resources, detrimentally impacting astronomy. A changing night sky with increased numbers of satellites also affects all those who rely on or merely enjoy the night sky.

With perhaps thousands of spacecraft predicted to be launched in the next decade, many of these headed to LEO, the challenge of protecting and preserving the night sky for science and society has taken the moniker Dark and Quiet Skies, and organizations including the International Astronomical Union (IAU) and its Centre for the Protection of Dark and Quiet Skies from Satellite Constellation Interference (CPS) have raised the salience of this issue in a variety of fora, including at COPUOS.

There are some technical approaches to minimize the harmful interference effects of these numerous satellites, and satellite operators have worked with the international astronomy community to reduce the brightness of their spacecraft. With the most severe interference threatened shortly after dusk, and again shortly before dawn, there are approaches to coordinating imaging times with the astronomy community. However, balancing the benefits of large global constellations in LEO with their detrimental impacts on astronomy appears to be a complex problem that a variety of stakeholders, including national regulators and international bodies, will continue to grapple with.

Status of Humans in Space

As states and private companies contemplate and prepare for crewed space operations ranging from suborbital to beyond Earth orbit, the legal status of humans in space within the international framework will need to be addressed. The treaty regime provides particular rights and responsibilities with regard to astronauts, and they may or may not apply to other spaceflight participants, such as private individuals and those flying on purely commercial flights.

Article V of the Outer Space Treaty refers to astronauts as "envoys of mankind," and requires that states give them "all possible assistance in the event of accident, distress, or emergency landing" in their territory or on the high seas. This assistance also requires their safe and prompt return to the state of registry of their space vehicle. In outer space and on celestial bodies, states must render "all possible assistance" to astronauts of other states party to the treaty. Last, states must also inform other states and the UN Secretary-General of any phenomena they discover in space that could constitute a danger to the life and health of astronauts.

The 1968 Agreement on the Rescue of Astronauts, the Return of Astronauts and the Return of Objects Launched into Outer Space further develops and refines the rights and obligations of humans in space. The plain meaning of these texts is largely clear, and highlights the peaceful and cooperative spirit animating the positive obligations it imposes upon states.

However, neither these treaties nor any subsequent source of international law defines the term "astronaut." It is likely that states seeking to build their space credentials will be interested in having their citizens join the list of humans who have traveled to outer space. As private spaceflight becomes a viable industry, and when suborbital spaceflight occurs across international boundaries, the status of private individuals and their recognition as astronauts will likely have to be reassessed on national and/or international levels.

Protecting Celestial Bodies

In addition to the environmental issues discussed in previous sections, the protection of celestial bodies is an advanced issue which some new actors in space may face. Article IX of the Outer Space Treaty first establishes a positive commitment where states shall be guided by the principle of cooperation and mutual assistance, and all activities shall be conducted with due regard for the

Figure 1.11 | NASA Astronaut Kate Rubins conducts a spacewalk on Sept. 1, 2016.
Credit: NASA

corresponding interests of other state parties. Concerning the environment of celestial bodies, all studies and exploration shall be pursued "as to avoid their harmful contamination."

The article then requires that states undertake "appropriate international consultations" before any activity or experiment they have reason to believe would cause potentially harmful interference with the space activities of other states. Last, states may request consultations concerning the activities or experiments of other states when they have reason to believe the activities or experiments would cause potentially harmful interference with their own activities.

While the text of the article is related to environmental protection, it is chiefly the second sentence of Article IX that concerns the protection of celestial bodies and creates the positive obligation for states to adopt appropriate measures to prevent the harmful contamination of outer space and celestial bodies. This sentence also concerns the creation of space debris and preventing the introduction of extraterrestrial matter to Earth. As such, this treaty article reflects the desire by states to preserve celestial bodies, and it has led to further elaboration on the meaning of planetary protection.

As mentioned, COSPAR has promulgated a Planetary Protection Policy for missions to other celestial bodies. Their Planetary Protection Policy, last revised in June 2020, reflects the concerns of scientists interested in the origin of life and the preoccupation that the environments of other celestial bodies might be contaminated, even unintentionally, by the arrival of crewed or robotic spacecraft. The updated Planetary Protection Policy lays out five categories of missions according to the destination involved and the type of mission (i.e., orbiter, lander, return-to-Earth mission; see Table 1.3).

Category I missions are those to celestial bodies lacking direct relevance for understanding the process of chemical evolution or the origin of life, and include certain types of asteroids and other destinations to be determined. No planetary protection concerns are defined for Category I missions, whether they be orbiters, rovers, or landers.

Category II missions also cover orbiters, rovers, and landers, but relate to missions to several major celestial bodies: Venus, Jupiter, Saturn, Uranus, and Neptune, as well as Ganymede, Callisto, Titan, Triton, Pluto and Charon, and Ceres, as well as comets, carbonaceous chondrite asteroids, and Kuiper Belt objects. These Category II missions address missions to celestial bodies where there is a significant scientific interest related to the process of chemical evolution or the origin of life, but, because of the physical environment of the destination, there is only a remote chance that contamination might compromise future investigations. Category II missions require a record of planned impact probability and contamination control measures, as well as a documentation of the planetary protection measures taken through the general planetary protection plan, a pre-launch report, post-launch report, post-encounter report, and an end-of-mission report.

	Category I	Category II	Category III	Category IV	Category V
Type of Mission	Any but Earth Return	Any but Earth Return	No direct contact (flyby, some orbiters)	Direct contact (lander, probe, some orbiters)	Earth Return
Target Body	See Category-specific listing	See Category-specific listing	See Category-specific listing	See Category-specific listing	See Category-specific listing
Degree of Concern	None	Record of planned impact probability and contamination control measures	Limit on impact probability Passive bioburden control	Limit on probability of non-nominal impact Limit on bioburden (active control)	If restricted Earth return: No impact on Earth or Moon; Returned hardware sterile; Containment of any sample.
Representative Range of Requirements		Documentation only (all brief): • PP plan • Pre-launch report • Post-launch report • Post-encounter report • End-of-mission report	Documentation only (Category II plus): • Contamination control • Organics inventory (as necessary) Implementing procedures such as: • Trajectory biasing • Cleanroom • Bioburden reduction (as necessary) • Contamination control • Organics inventory (as necessary) Implementing procedures such as: • Trajectory biasing • Cleanroom • Bioburden reduction (as necessary)	Documentation (Category II plus): • Pc analysis plan • Microbial reduction plan • Microbial assay plan • Organics inventory Implementing procedures such as: • Trajectory biasing • Cleanroom • Bioburden reduction Partial sterilization of contacting hardware (as necessary) • Bioshield • Monitoring of bioburden via bioassay	Outbound Same category as target body/ outbound mission Inbound If restricted Earth return Documentation (Category II plus): • Pc analysis plan • Microbial reduction plan • Microbial assay plan Implementing procedures such as: • Trajectory biasing • Sterile or contained returned hardware • Continual monitoring of project activities • Project advanced studies and research If unrestricted Earth return: None

Table 1.3 | This table from COSPAR shows the level of contamination concern for each COSPAR mission type. Additionally, the chart shows a short list of requirements that would need to be fulfilled to meet these protection standards. *Credit: COSPAR Planetary Protection Policy*

orbiters to Mars, Europa, or Enceladus (Category III), landers to Mars, Europa, or Enceladus (IV), or any Earth-Return mission (V). Earth-Return missions from Venus or the Moon are classified as "Unrestricted Earth Return," while missions to and from Mars or Europa are "Restricted Earth-Return Missions" requiring heightened scrutiny.

COSPAR guidelines are implemented on a national level, where space agencies and governments adopt them into their national licensing and regulatory frameworks or implement them in the national space agency's plans. In the United States, NASA has a Planetary Protection Office, an agency-wide policy directive, and mandatory procedural requirements for its missions. ESA also has a planetary protection policy (ESSB-ST-U-001 (Issue 1): ESA planetary protection requirements), which it imposes on all ESA spaceflight missions, any contributions to ESA missions, and any ESA contributions to non-ESA missions.

Outside of planned missions to preserve celestial bodies for their scientific value, there is also the desire to protect and preserve certain areas and artifacts on celestial bodies because of their importance to space exploration. The landing sites of the Apollo missions, including the hardware the astronauts left on the Moon and even the iconic footprints from Neil Armstrong, Buzz Aldrin, and later astronauts are of permanent cultural value. The same is true for the Soviet-era rovers on the surface of the Moon, such as Lunokhod, and of the rovers on other celestial bodies. While there has been talk of making some of these sites United Nations Educational, Scientific and Cultural Organization (UNESCO) World Heritage sites before they are encroached upon by next-generation missions, so far it is up to national governments and their space agencies to try to preserve those sites and artifacts in which they have particular interest. For missions planned to certain destinations, responsible actors would be well advised to educate themselves on the various planetary protection policies and to observe them in the execution of their missions. This is especially important if national legislation is silent on the topic of planetary protection.

Space Resources

As discussed at the beginning of this chapter, there are significant freedoms to explore and use outer space. The Outer Space Treaty even ensures that this exploration and use is "the province of all mankind." But what rights do states, private companies, or even people have to use space resources, and what does the term "space resources" encompass? While the drafters and negotiators of the Outer Space Treaty considered this topic, they left it vague enough to allow further refinement. However, this treaty—while enshrining significant freedoms in space— does have some prohibitions. Article II of the Outer Space Treaty states that:

> Outer Space, including the Moon and other celestial bodies, is not subject to national appropriation by claim of sovereignty, by means of use or occupation, or by any other means.

By referencing the Moon and other celestial bodies in this article, it is very clear that the prohibition on national appropriation applies to both physical celestial bodies and to "void" space. More importantly, the listing of claims of sovereignty, and the use of or the occupation of space, is a list of methods (or means) that would not justify a state's appropriation of outer space. Neither a statement (such as a claim) nor a physical act (such as using or occupying) constitutes lawful appropriation. Article II's list is not exhaustive: it is merely illustrative of a few explicit methods that will not legitimize national appropriation in space.

Figure 1.12 | Internal asteroid contents from Japan's Hayabusa2 mission, which collected samples from asteroid 162173 Ryugu.

Credit: JAXA

The further interpretation and possible clarification of Article II of the Outer Space Treaty will likely become necessary as a range of space resources activities expand. Notably, China, the United States, and the USSR have already brought back lunar samples, and have acted in ways consistent with asserting and transferring uncontested ownership rights in those samples, and the United States and Japan have also conducted missions that returned samples collected from asteroids.

While the issue of the use of space resources is being examined, the purposes of the Outer Space Treaty would seem counter to any overly drastic prohibitions that would limit the next generation of space activities. As long as the use of space resources conforms to the purposes of the treaty, advances the aims of the treaty, and otherwise conforms to international law, it is permissible. Additionally, as long as these activities do not rise to the level of a state attempting the national territorial annexation of outer space or celestial bodies, they are likewise permissible.

In 2019, the Hague International Space Resources Governance Working Group (a consortium of industry and academia) finalized their *Building Blocks for the Development of an International Framework on Space Resource Activities*. The Hague Building Blocks demonstrate the further elaboration of elements of normative regime for the use of space resources, the definition of the term space resources, and have influenced national and international thought on the issue.

Meanwhile at COPUOS, a Working Group on Legal Aspects of Space Resource Activities was established in 2021. The Working Group has a Terms of Reference, five-year Workplan, and Method of Work on potential legal models for regulating space resources.

Additionally, national approaches and multilateral agreements like the Artemis Accords signal how states are approaching advanced space governance issues. See Chapter Two for more information on this example.

Future of International Space Governance

Another topic that new space actors must be aware of is the changing nature of international space governance. This chapter has mentioned and discussed the traditional institutions and committees where space governance was historically developed. These include COPUOS, as the home of the UN treaties on outer space, a venue that continues to serve as a "clearinghouse" of information on national space exploration activities, and as a venue for deliberations on new rules. Most recently, COPUOS adopted Guidelines for the Long-term Sustainability of Outer Space Activities, and the Working Group on the Long-term Sustainability of Outer Space Activities continues to meet, along with discussions under a variety of other agenda items touching on space sustainability.

However, other important multilateral fora discuss space security and disarmament issues in a variety of venues at the UN in Geneva, such as the UN Disarmament Commission, the Conference on Disarmament, events put on by UNIDIR (the United Nations Institute for Disarmament Research), and in various UN Group of Governmental Experts (GGE) and UN Open-Ended Working Groups (OEWGs) considering various issues on space security.

The ITU has also served as a venue for the development of norms related to outer space, including in study groups and at its World Radiocommunication Conferences. This and the following chapters show that rules regulating behavior on space have also been developed outside the UN system, including in regional organizations such as the North Atlantic Treaty Organization (NATO), and groups of states cooperating on space exploration goals such as through the Artemis Accords and the International Lunar Research Station (ILRS) agreements.

Meanwhile, rules have been developed at the national level through the creation of national space policy, and in national space legislation and its implementing regulations. The commercial and scientific sectors have also self-organized to develop standards and recommended practices for their own constituencies. See Chapter Two and Chapter Three respectively for a deeper discussion of these issues.

This multifaceted and diverse nature of rule development for outer space activities is a feature of global space activities, and an awareness of this "polycentric" nature of globalized space governance will benefit both new and traditional space actors.

CHAPTER ONE ADDITIONAL RESOURCES

See below for additional resources, links, and documents referenced in Chapter One.

Building Blocks for the Development of an International Framework on Space Resource Activities: https://www.universiteitleiden.nl/binaries/content/assets/rechtsgeleerdheid/instituut-voor-publiekrecht/lucht--en-ruimterecht/space-resources/bb-thissrwg--cover.pdf

Committee on Earth Observation Satellites: https://ceos.org/

Convention on Registration of Objects Launched into Outer Space: https://www.unoosa.org/oosa/en/ourwork/spacelaw/treaties/registration-convention.html

COPUOS Legal Subcommittee: Space Resources: https://www.unoosa.org/oosa/en/ourwork/copuos/lsc/space-resources/index.html

COSPAR Planetary Protection Policy: https://cosparhq.cnes.fr/assets/uploads/2019/12/PPPolicyDecember-2017.pdf

EU Space Surveillance and Tracking: https://www.eusst.eu/

Group on Earth Observations: https://earthobservations.org/

Guidelines for the Long-term Sustainability of Outer Space Activities of the Committee on the Peaceful Uses of Outer Space: https://www.unoosa.org/oosa/en/oosadoc/data/documents/2021/stspace/stspace79_0.html

Hague Code of Conduct against Ballistic Missile Proliferation: https://www.hcoc.at/

Inter-Agency Space Debris Coordination Committee: https://iadc-home.org/what_iadc

IADC Space Debris Mitigation Guidelines: https://iadc-home.org/documents_public/file_down/id/5249

IADC Statement on Active Debris Removal: https://newspaceeconomy.ca/wp-content/uploads/2023/05/iadc_statement_on_active_debris_removal.pdf

IAEA Safety Framework for Nuclear Power Source Applications in Outer Space: https://www.iaea.org/sites/default/files/safetyframework1009.pdf

IAU Centre for the Protection of Dark and Quiet Skies from Satellite Constellation Interference (CPS): https://cps.iau.org/

International Astronomical Union: https://www.iau.org/

International Charter Space and Major Disasters: https://disasterscharter.org/web/guest/about-the-charter

International Committee on Space Research: https://cosparhq.cnes.fr/

International Court of Justice: https://www.icj-cij.org/home

International Telecommunication Union (ITU): https://www.itu.int/en/Pages/default.aspx

ITU Master International Frequency Register: https://www.itu.int/en/ITU-R/terrestrial/broadcast/Pages/MIFR.aspx

ITU Telecommunication Standardization Sector: https://www.itu.int/en/ITU-T/Pages/default.aspx

ITU World Radiocommunication Conferences: https://www.itu.int/en/ITU-R/conferences/wrc/Pages/default.aspx

Manual on International Law Applicable to Military Uses of Outer Space: https://www.mcgill.ca/milamos/

Missile Technology Control Regime: https://www.mtcr.info/en

NASA Planetary Protection Office: https://sma.nasa.gov/sma-disciplines/planetary-protection

National Registers of Space Objects: https://www.unoosa.org/oosa/en/spaceobjectregister/national-registries/index.html

Space Data Association: https://www.space-data.org/sda/

Space Debris by the Numbers: https://www.esa.int/Space_Safety/Space_Debris/Space_debris_by_the_numbers

Space Debris Mitigation Guidelines of the Committee on the Peaceful Uses of Outer Space: https://www.unoosa.org/documents/pdf/psa/bsti/COPUOS_SPACE_DEBRIS_MITIGATION_GUIDELINES.pdf

Space Frequency Coordination Group: https://www.sfcgonline.org/home.aspx

U.S. Department of Commerce Space Industry Technical Standards Compendium: https://www.space.commerce.gov/space-industry-technical-standards/

UN UN Inter-Agency Space Debris Coordination Committee: https://www.unoosa.org/oosa/en/ourwork/icg/icg.html

UN Outer Space Objects Index: https://www.unoosa.org/oosa/osoindex/index.jspx?lf_id=

UN Resolution 1962 (XVIII)—(Declaration of Legal Principles Governing the Activities of States in the Exploration and Use of Outer Space): https//www.unoosa.org/oosa/en/ourwork/spacelaw/principles/legal-principles.html

UN Resolution 2222 (XXI)—Outer Space Treaty (Treaty on Principles Governing the Activities of States in the Exploration and Use of Outer Space, including the Moon and Other Celestial Bodies): https://www.unoosa.org/oosa/oosadoc/data/resolutions/1966/general_assembly_21st_session/res_2222_xxi.html

UN Resolution 2345 (XXII)—Astronaut Agreement (Agreement on the Rescue of Astronauts, the Return of Astronauts and the Return of Objects Launched into Outer Space): https://www.unoosa.org/oosa/oosadoc/data/resolutions/1967/general_assembly_22nd_session/res_2345_xxii.html

UN Resolution 2777 (XXVI)—Liability Convention (Convention on International Liability for Damage Caused by Space Objects): https://www.unoosa.org/oosa/oosadoc/data/resolutions/1971/general_assembly_26th_session/res_2777_xxvi.html

UN Resolution 3235 (XXIX)—Registration Convention (Convention on Registration of Objects Launched into Outer Space): https://www.unoosa.org/oosa/oosadoc/data/resolutions/1974/general_assembly_29th_session/res_3235_xxix.html

UN Resolution 34/68—Moon Agreement (Agreement Governing the Activities of States on the Moon and Other Celestial Bodies): https://www.unoosa.org/oosa/oosadoc/data/resolutions/1979/general_assembly_34th_session/res_3468.html

UN Resolution 37/92—(Principles Governing the Use by States of Artificial Earth Satellites for International Direct Television Broadcasting): https//www.unoosa.org/oosa/en/ourwork/spacelaw/principles/dbs-principles.html

UN Resolution 41/65—(Principles relating to remote sensing of the Earth from outer space): https//www.unoosa.org/oosa/en/ourwork/spacelaw/principles/remote-sensing-principles.html

UN Resolution 47/68—(Principles Relevant to the Use of Nuclear Power Sources in Outer Space): https//www.unoosa.org/oosa/en/ourwork/spacelaw/principles/nps-principles.html

UN Resolution 55/122—(International Cooperation in the Peaceful Uses of Outer Space): https//www.unoosa.org/oosa/oosadoc/data/resolutions/2000/general_assembly_55th_session/ares55122.html

UN Resolution 59/115—(Application of the concept of the "launching State"): https//www.unoosa.org/oosa/oosadoc/data/resolutions/2004/general_assembly_59th_session/ares59115.html

UN Resolution 62/101—(Recommendations on enhancing the practice of States and international intergovernmental organizations in registering space objects): https//www.unoosa.org/oosa/oosadoc/data/resolutions/2007/general_assembly_62nd_session/ares62101.html

UN Resolution 68/74—(Recommendations on national legislation relevant to the peaceful exploration and use of outer space): https//www.unoosa.org/pdf/gares/A_RES_68_074E.pdf

UN Resolution RES 1721 (XVI)—(International Co-operation in the Peaceful Uses of Outer Space): https//www.unoosa.org/oosa/en/ourwork/spacelaw/treaties/resolutions/res_16_1721.html

UN Resolution–Space Benefits Declaration: https://www.unoosa.org/oosa/en/ourwork/spacelaw/principles/space-benefits-declaration.html

United Nations Committee of Experts on Global Geospatial Information Management: https://ggim.un.org/

Wassenaar Arrangement: https://www.wassenaar.org/

Woomera Manual on the International Law of Military Space Activities and Operations: https://law.adelaide.edu.au/woomera/

World Meteorological Organization Resolution 40: https://community.wmo.int/en/resolution-40

CHAPTER TWO

National Space Policy and Administration

CHAPTER TWO FOCUS

This chapter provides an overview of how and why states create national frameworks for space activities through policy, regulation, and the organizations they create to implement it. A policy is a principle or a set of principles used to guide decision-making and actions.

In the context of governments, "public policy" refers to why, how, and to what effect governments pursue particular courses of action or inaction. Public policy decisions often involve weighing the potential positive and negative impacts of competing options. These decisions are further complicated by the participation of many different interest groups and political actors who have competing perspectives in the decision-making process. In conjunction, "public administration" is the implementation of policy through the organization of government bureaucracy, the establishment of programs and institutions, and the day-to-day running of services and activities.

This chapter is divided into two main sections. The first section focuses on public policy aspects of national frameworks, including various ways space policy can be established; why states put in place national policy; the relationship between space and science, technology, and innovation policy; and the role of international cooperation. The second section focuses on public administration: how countries implement their own national policy and international obligations through regulative and administrative structures.

Dr. Joel Joseph S. Marciano Jr.
DIRECTOR GENERAL
Philippine Space Agency (PhilSA)

INTRODUCTION

A new space actor faces many challenges. Apart from building up essential science and getting the technology right, venturing into space also requires a host of policies, regulations, and decisions at the national level. For the Philippines, all of these had to come together to capture and convey our goals and direction when we started the Philippine Space Agency (PhilSA) in 2019.

The Philippine Space Policy, established under the Philippine Space Act (Republic Act No. 11363) signed by then Philippine President Rodrigo R. Duterte, addresses, among others, the "urgent need to create a coherent and unified strategy for space development and utilization to keep up with other nations in terms of space science and technology". PhilSA was also created under the same Act and given the mandate to plan, develop, and promote the national space program in line with the Philippine Space Policy. As the central agency of the Philippine government addressing all national issues and activities related to space science and technology applications, the Agency is placed directly under the Office of the President of the Philippines for policy and program coordination and to ensure alignment in national policies and priorities.

New space actors need guidance and rubric from national space policy in the process of establishing themselves. Governments have responsibilities under the Outer Space Treaty to provide authorization and continuing supervision over their national actors, and cascading those responsibilities into national policy is up to each country. In the Philippines, we are taking an end-to-end and holistic approach to building the domestic space ecosystem that unlocks important socioeconomic benefits for the people.

Space is important for my country. We leverage satellites for communication, weather monitoring, hazard management, climate studies, environmental protection, maritime safety, and other important uses that bolster national security and the attainment and protection of national interest – seeking constructive dialogue and cooperation with other countries in the process. We put in place domestic policy that enshrines access to space and its environs as a sovereign right, and to enable us to secure opportunities for creating, adding, and deriving value from such access. The work continues.

Amid the challenges confronting new actors, a responsive policy and regulatory framework can provide encouragement and the enabling environment for them to venture into space in a safe, responsible and sustainable way. Thus, in this chapter, different elements of national space policy are discussed, diving into issues such as public policy for space, the role of space in science and technology policy, export control considerations, and remote sensing regulations. After reading through it, new actors in space should have a better understanding of the types of regulations they might need in support of their burgeoning space dreams and ambitions.

CHAPTER TWO
National Space Policy and Administration

PUBLIC POLICY FOR SPACE

Policy can be established through many different methods, several of which may be interacting at the same time. One way of establishing policy is through the international, bilateral, and multilateral treaties and agreements by which a state is bound. National policy can be established explicitly through formal decision-making processes such as intra-governmental committees or legislation. Policy can also be established implicitly through a choice to not pursue a particular path and can be manifested through cultural or ideological contexts that impact decision-making and choices. Policy may not be consistent and may even be contradictory.

In the context of space, policy can take many different forms. Some states choose to put in place a national space policy, which may or may not be accompanied by narrower policies covering specific space sectors such as launch, communications, or remote sensing. Other states choose to put in place policy at the organizational level, or through legislation that establishes specific programs and projects. Still others express it through a strategy that spells out national goals and priorities for space. Making national space policy or strategy publicly accessible is one way to demonstrate intentions and priorities for a national space program. It also gives an idea of how much budgeting may go into a nation's space activities and raises the overall level of transparency. In addition, developing a national space policy or strategy forces a government to go through the process

> Developing a national space policy or strategy forces a government to go through the process of having an intragovernmental discussion about priorities and goals for its space program, information which can then be used to inform national and international discussions.

of having an intragovernmental discussion about priorities and goals for its space program, information which can then be used to inform national and international discussions. The Handbook for Space Capability Development is another resource that looks at this topic of how governments go about drafting a national space strategy and building a national space program. The following sections provide an overview of the different uses and common elements of space policy.

Rationales, Objectives, and Principles

A national space policy provides the rationale for why a state chooses to engage in space activities. The reasoning and motivation for engaging in space activities may differ drastically between states. Some states choose to engage in the entire spectrum of space activities and capabilities across the commercial, civil, and national security sectors, while other states choose to focus on or exclude specific types of activities. In some cases, this choice may reflect a national decision on a specific interpretation of what the peaceful uses of space means, or a state's relationship and ideological approach to its private sector. Explicitly and publicly defining the rationales for space activities may also be part of a strategy for boosting internal political support for funding and other resources which support space activities.

National space policy also provides the objectives for the space activities a state chooses to engage in. The reason for doing so is to provide high-level guidance on the goals a state is pursuing. These goals can be specific, such as accomplishing a certain task in a set amount of time, or broad, such as enhancing national prestige or bulwarking national security goals. Explicitly outlining these objectives not only provides a signal to other countries, but also can help generate national support and motivation for specific space activities and programs.

National space policy can also define the principles by which a state will conduct its space activities. These principles can be used to reaffirm or demonstrate a state's adherence to international agreements and treaties, and to outline national principles that have a historical, cultural, or ideological basis. The principles in a national space policy can also form the foundation for lower-level government policies in specific sectors such as national security or commercial space.

When a new space activity is proposed, its alignment with national policy and principles related to space can be an early indicator of whether that activity will be implemented successfully. If serious misalignments exist, strategies for overcoming those misalignments need to be addressed in the planning process.

Government Roles and Responsibilities

A second major use for national space policy is to delineate roles and responsibilities between various government agencies and entities to comply with a state's obligations under the international legal framework for space activities discussed in Chapter One. States need to assign responsibility to government entities performing functions such as administering and licensing radio frequencies used by satellites, licensing remote sensing satellites, and maintaining a national registry of space objects.

States have multiple options for how to assign roles and responsibilities. Some countries choose to consolidate all of their space activities into one organization, while others have multiple government entities that are each tasked with a portion

Case Study: United Arab Emirates Mars Mission

In July 2014, the government of the United Arab Emirates (UAE) announced its intentions to develop and launch a robotic spacecraft to Mars orbit. The plan marked an ambitious expansion of the UAE's space activities, which had previously focused on remote sensing and communications and coincided with the establishment of the United Arab Emirates Space Agency. The UAE's commitment to a scientific Mars exploration project encompasses many of the typical goals and drivers that are found in government space programs.

Emirati officials described three key motivations for the project: symbolism and inspiration; acting as a catalyst for knowledge and skill development; and providing an anchor project for the domestic space industry in the UAE. The timing of the spacecraft launch was symbolically important, as it arrived at Mars in 2021 to coincide with the 50th anniversary of UAE independence. The mission was also named "Hope" with the explicit purpose of sending a message of optimism. The UAE defined specific science objectives for the mission, and involved local universities in the execution of scientific activities. It also planned for the spacecraft and associated mission support elements to be manufactured entirely by Emirati citizens, with up to 150 people employed in the program.

While Emirati-led, the program also demonstrated the role international partnerships often play in the execution of national space programs. The spacecraft was launched on a Japanese launch vehicle, and the government of the UAE entered into several cooperative agreements with other nations (including the United States and Russia) to exchange information related to Mars science and exploration. Through these agreements, the UAE sought access to training and knowledge development for its scientists and engineers. To that end, the UAE Space Agency also entered into an agreement with Lockheed Martin under which a training program in space-related skills was established for students and young professionals. Although not solely related to the Mars mission, this program demonstrated the UAE's emphasis on linking space development to scientific and technical capacity-building.

Hope successfully arrived in Mars orbit as planned in February 2021, making the UAE the first Arab country and the fifth country to reach Mars. Hope has provided significant scientific returns, including insight into the Martian atmosphere and the first-ever close up global images of the Martian moon Deimos.

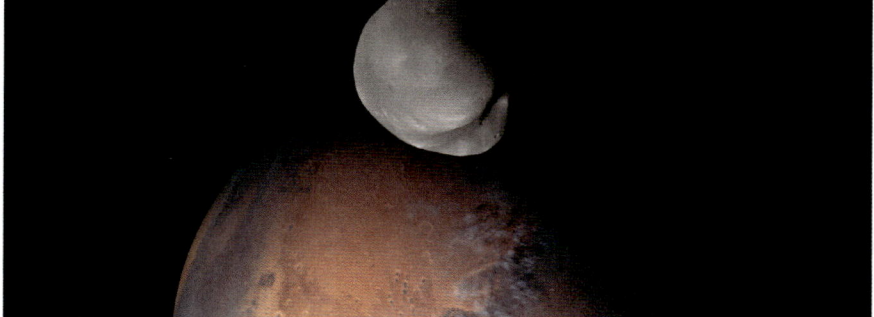

Figure 2.1 | A digital composite of Mars with its smaller moon, Deimos, combining data captured by the Hope Orbiter from the Emirates Mars Mission. *Credit: United Arab Emirates Mars Mission*

of the space activities or oversight. This division of labor could be functional, such as dividing licensing responsibilities between agencies depending on their expertise. The division could also be between civil and national security space activities, in order to enable easier public acknowledgment and international cooperation while also protecting sensitive technology or capabilities.

National space policy can also be used to direct coordination between national agencies or entities. If roles and responsibilities are divided among multiple government agencies, it is often the case that there will be a need for some of those agencies to coordinate their activities with other entities. This coordination may not happen naturally, as it can involve disputes over power, control, and budget. Space policy can be used to direct coordination with other agencies in situations where their responsibilities overlap, or direct coordination with private sector or international entities to accomplish policy objectives and principles.

The process by which a government makes national space policy decisions is important and can vary widely by country. The intra-governmental decision-making process helps ensure that space related policies are consistent with larger policy objectives, such as foreign policy or innovation policy objectives. Decisions that are made by individual government agencies or entities without coordination and input from other stakeholders, including the private sector, are likely to be suboptimal, because barriers between commercial, civil, and national security space activities are becoming increasingly blurred. Most space technology is dual-use, in that it can have both military/security and commercial/civilian applications which can be carried out simultaneously or alternately. Therefore, policy decisions on space technology need to strike a balance between controlling access to the technology to minimize national security risks and increasing access to maximize its socioeconomic benefits. As a result, policy decisions related to space activities will often result from coordination and collaboration among the relevant government agencies and bodies, and may benefit from the input of advisory bodies that represent other stakeholders, both within and outside of government.

It is important to have administrative and regulatory capacity for implementation of national policy and associated responsibilities. New state actors should determine how best to implement their international obligations, particularly their Outer Space Treaty Article VI responsibilities to authorize and provide continuing supervision for their national space activities, while also advancing their domestic priorities.

Although each state's national space policy is a unique reflection of its politics, culture, and priorities, there are a few common themes that occur across many national space policies. These themes reflect common challenges that states face and priorities as they try to promote through their national space policies.

Role of Space in Science, Technology, and Innovation Policy

The significant socioeconomic benefits reaped by established space nations from their space activities have been cited as a key motivator by emerging space countries making initial investments in space. Often tied to larger strategic goals for national science, technology, and innovation (STI) policy, space activities may include a high degree of investment in basic science and in research and development (R&D), with the goal of contributing to the national economy in sectors other than space. In this respect, a government's space policy may be a subset of STI policy, and space may be one of several target innovation areas, such as energy, aeronautics, public health, and computing.

STI policies will generally focus on the interactions among the relevant government, academic, and industry actors involved in education, basic and applied science, technology, and innovation. The coordination of STI-related efforts among the different actors is often a key challenge, as is the ability of actors within the ecosystem to integrate innovative products or processes. One particular challenge is overcoming the gap between moving from basic research to commercialization, sometimes referred to as the "valley of death." In this respect, STI policies will seek to not only incentivize innovation (e.g., intellectual property rules, competitive grants or awards), but also to develop the mechanisms to sustain innovation through the different development cycles so it can yield the desired economic gains.

As an example, the government of India adopted the Indian Space Policy in 2023 in order to give more regulatory stability to its private sector with the hope of expanding that part of its economy and using space as a driver for STI. Under this policy, most of the restrictions on commercial participation in India's space sector have been removed. The Indian Space Research Organisation (ISRO) will be responsible for research and development of new space technology and applications, including human spaceflight, and the Indian National Space Promotion Authorization Center (IN-SPACe) will authorize and regulate space activities, with the goal of releasing guidelines and procedures that would facilitate business for commercial space actors.

> The significant socioeconomic benefits reaped by established space nations from their space activities have been cited as a key motivator by emerging space countries making initial investments in space.

Among the primary goals often contained in STI policies is the development of a highly skilled workforce through investments in science, technology, engineering, and mathematics (STEM) education. The development of human capital is considered fundamental in industrial policy as part of efforts to develop niche capabilities and reduce the emigration of skilled or highly educated workers, also known as "brain drain." For example, Argentina's National Space Plan 2016-2027, developed by CONAE (Comisión Nacional de Actividades Espaciales), mandates that it must "contribute to the national scientific-technological field, both advanced knowledge and new educational and work opportunities through the creation of specific careers and related specialties." This motivation is also reflected in the practice of many countries seeking partnerships that include capacity-building components as a way to build human capital and grow national technological capacities.

Placing space activities within a larger STI framework can help answer critical questions about the long-term goals of these activities, how they relate to other science and technological efforts, and how best to coordinate among government and non-governmental efforts.

International Cooperation

International space cooperation is a key aspect of most space programs. Depending on the objectives, this cooperation can take many forms, such as multilateral cooperation at the international or regional level and bilateral cooperation with individual countries. Depending on the format of this cooperation, countries may designate specific agencies or institutions as the main representative, but the activity may involve other agencies or departments and non-governmental representatives from industry or academia.

At the multilateral level, active participation in the key space forums such as the United Nations Committee on the Peaceful Uses of Outer Space or the International Telecommunication Union, as well as related forums for cooperation in specific application areas such as the Group on Earth Observations, is often considered a fundamental aspect of these activities. Countries see it as both a way to show leadership and ensure their views are represented in relevant exchanges at the international level, while also sharing information about their space activities and learning about the activities of others. This participation may thus influence policy debates at the national level.

At a regional or bilateral level, countries may adopt multiple mechanisms to formalize relationships—whether through issuing joint declarations or statements, signing cooperative agreements to pursue specific activities together or to exchange data, pooling institutional or financial resources in a cooperative program, or by other methods. Regional space cooperation organizations have also emerged as a way to improve cooperation in and coordination of space activities at the regional level. For example, the Asia-Pacific Regional Space Agency Forum (APRSAF) seeks to advance space activities in the Asia-Pacific region with institutions from more than forty countries participating along with private sector entities.

While an exhaustive description of the multiple mechanisms actors have pursued to enable international cooperation is beyond the scope of this section, the key insight is that international cooperation is rarely pursued haphazardly, but is instead often part of larger policy and strategic considerations. International cooperation is often considered both a mechanism and a goal, so it may feature in policy documents. As a mechanism, space cooperation enables actors to leverage the expertise, investments, and resources of others in the development of programs, whether through the direct acquisition of hardware or the joint development of technical capacity.

International cooperation can also be driven by larger policy objectives and be part of a strategy to advance foreign policy, innovation, or trade policy goals. In emerging space countries, the two aspects may be tightly linked. Chile's space

Case Study: Lunar Governance Initiatives

There has been a renewed global interest in lunar and cislunar space activities, with more than hundred robotic and human exploration missions planned over the next decade. At the same time, there is also a renewed discussion of how existing space governance frameworks and principles apply to these new uses, users, and activities on or near the Moon. While some of these discussions are being held in multilateral fora such as the United Nations Committee on the Peaceful Uses of Outer Space (UN COPUOS), groups of states have also created their own initiatives.

The Artemis Accords are a set of principles for lunar activities that was initiated by the United States and first announced in October 2020 with the signing by eight initial countries. The Artemis Accords are related to the Artemis program, a NASA-led initiative to return to the Moon and establish a permanent human presence there that lays the foundation to further exploration to Mars and beyond. NASA and the U.S. State Department are co-leads for the Artemis Accords.

The Artemis Accords build on the principles contained in the Outer Space Treaty of 1967 and apply them to the lunar space activities. The Artemis Accords' principles address transparency, interoperability, release of scientific data, resource utilization, and more. In 2023, the Artemis Accords partners started a series of working groups to discuss the specifics of how the principles in the Accords will be applied to their future lunar activities.

Another example of states starting their own lunar initiatives is the International Lunar Research Station (ILRS), which is led by China and Russia and was announced in June 2021. Similar to Artemis, the ILRS consists of both a lunar exploration program and a set of principles for activities undertaken as part of that program.

In June 2021, China and Russia released the "ILRS Guide for Partnership" that provides details about the program's scientific objectives, mission phases, and guidelines for partnership. It outlines the Joint Working Group that will oversee the legal, scientific, and engineering aspects of ILRS. In 2023, China further described the intended creation of an International Lunar Research Station Cooperation Organization (ILRSCO) that would handle the cooperative aspects of the program.

Currently, no details about the ILRS principles are publicly available. It is unclear if signing the ILRS principles will be a prerequisite for participation in the ILRS program and vice versa. It is also unclear if the ILRS principles will differ significantly from those contained in the Artemis Accords. However, the similarities between the activities planned under Artemis and ILRS (permanent installations, extraction and use of lunar water and mineral resources, and manufacturing on the lunar surface) suggests they will be similar.

Figure 2.2 | With NASA Administrator Bill Nelson and Netherlands Ambassador to the United States Birgitta Tazelaar looking on, Harm van de Wetering, Director of the Netherlands Space Office, signs the Artemis Accords. *Credit: NASA*

policy, for example, is designed with the goal of positioning it as a hub for services and scientific-technical development so it can be a relevant actor in the region. For Chile and other countries in the Latin America and Caribbean region, international cooperation—particularly bilateral and regional cooperation—is considered a priority as a way to extend limited resources, as well as to support related strategic and political goals. Often states with established space capabilities may pursue international space cooperation, not because of resource limitations, but as an added measure to foster positive relationships with other countries.

In this respect, national space policies may detail the goals and priorities of international cooperation efforts, a mechanism that helps signal others about a government's priorities and goals in space, enhances transparency of their activities with partner nations, and invites new actors to identify opportunities for engagement.

Export Control and Technology Transfer

The underlying question when working on export controls is, with the increased access to space and burgeoning role of the private sector in space, how does a state balance controlling the proliferation of militarily sensitive technologies with commercial development and innovation? It is particularly challenging to do so while supporting and propelling the space industrial base—an objective of many national space strategies—as export control is perceived to be a necessary part of ensuring national security and assuring a stable and predictable space environment. The balance between efficiency and commercial interests on one hand and national security on the other is a difficult one to strike; another way of looking at this is as being part of a larger discussion about promoting innovation while minimizing risks.

Keeping in mind the international aspect of export control discussed in Chapter One: International Export Control, export control restrictions on the national level are extremely challenging to implement and, as a result, should be undertaken only after a considerable amount of discussion with all stakeholders, including industry, and when the government has a solid understanding of what it is trying to accomplish with export control protections. Without stakeholder input, the domestic industry can suffer unduly with very little benefit to a country's national security. States have to be careful of unintended consequences: for example, cases where export controls were changed and thus created new burdens for groups in the space industry, or where they unduly hampered domestic industry even though the same technologies or capabilities were widely available internationally. It is important to get the conversation as wide as possible when creating or updating government regulation of an industry, and to have an open conversation with industry to ensure that all aspects of an issue are considered.

Maintaining a list of technologies that should be controlled is challenging, particularly for space technologies, many of which are dual-use. One sticking point for export controls is that often the technology outpaces the legal regimes. How helpful are export regulations when they are essentially protecting outdated technology? Another significant issue is that export control, by its nature, tries to control the technology or goods themselves, regardless of how they are being used. This runs contrary to one of the emerging lessons from dealing with dual-use space technologies: it is more important and effective to focus on the actions and uses than on the technology itself.

Balancing national security concerns and positions on fostering domestic industries and innovation is often an early and important consideration for new state actors. For non-governmental actors, a thorough understanding of relevant export control regimes must begin early in the planning process.

Government Relationship with the Private Sector

Governments occupy a range of roles in their interaction with the private sector: regulator; customer; supplier of technology, intellectual property, and/or seed funding; collaborator; and competitor. The way these roles are expressed is a major influence on the development of a broader space industry outside of the government program in a given country. Along with its role in the market as a regulator, the government also exerts considerable influence through its role as a customer. Governments must be aware of how the choices they make in engaging the private sector through the procurement of goods and services affects both the development of industry and the evolution of government space strategy and programs.

Governments may choose to develop required capabilities or services internally, and not engage the private sector at all. There are several scenarios in which this approach may be preferred: the capability may not exist in the private sector, a determination might have been made that development of the capability is considered a core governmental function (for example, a capability used for national security purposes), or the capability provides a public good. Developing capabilities internally to the government provides the government with complete control over execution of the project as well as any intellectual property developed. It may help the government remain abreast of current technology, and can help government personnel stay engaged with program execution. However, in-house work has drawbacks, including a lack of transparency, and potential cost and efficiency challenges as compared to wholly private work. As well, governments need to be able to balance developing a capability in-house while also promoting a related commercial sector (if one is wanted for that capability). Governments must also remain aware of similar capabilities that the private sector may be developing in order to ensure that approaches remain current with regard to comparable capabilities.

> Governments occupy a range of roles in their interaction with the private sector: regulator; customer; supplier of technology, intellectual property, and/or seed funding; collaborator; and competitor.

By contracting required capabilities out to the private sector, the government is able to foster private sector capacity and expertise, which in theory supports broader economic development objectives. Competition may also lead to more innovative solutions than might be developed if the work were to be completed in house. In general, contracting with the private sector is intended to provide capabilities in a more cost-effective and efficient manner than the developing of capabilities internally to the government. However, contracting imposes administrative costs on both the government and the private-sector entities, specifically in terms of administration and performance oversight. While contracts can provide the government with a certain level of oversight and ability to specify quality level, contract performance attributes, and execution timelines, it should be acknowledged that the contracting process inherently involves a decision to cede some control of the development of the capability. Contracting may also create dependencies between the government and companies receiving contracts. The government may find itself dependent upon one or a few suppliers for a critical capability, and companies may find themselves dependent upon the government as a critical source of revenue. Companies may go bankrupt, leaving the government without the specific capability that they provided.

Due in part to these drawbacks, governments are increasingly utilizing public-private partnership-based approaches to engage the private sector. Public-private partnership approaches typically seek to develop capabilities in a way that ensures both the government and the participating private sector entities are co-invested in the success of the activity. Commonly, governments might specify a need and some basic requirements, as well as allocate a certain amount of funding. The capability to be acquired is one that the commercial sector can use to satisfy non-governmental requirements, with the governmental funding intended to be complemented by investment and capital provided by the commercial sector. Projects of this type give the government less control over the execution of the project but can provide capabilities at less cost than traditional contracting. The private-sector participants are required to invest their own funding. However, they are able to retain ownership of the products and intellectual property produced. These types of activities may also be used to stimulate the development of capabilities that require governmental support to overcome initial research and development costs.

Governments may also procure capabilities on a purely commercial basis, often referred to as commercial-off-the-shelf (COTS). In this approach, the private sector offers items at a standard price, commonly via a catalog. Governments are able to purchase those items in a market transaction no different from business-to-business sales. These sorts of transactions have a lower administrative burden than contracting approaches. They are typically used for the purchase of bulk goods or commodities. The government is able to procure required items quickly and efficiently but is not able to specify the details of the development process.

Governments may choose to acquire capabilities through the use of grants instead of contracts. Grants are typically used in situations where the government's interest is in acquiring research or technology development

Case Study: Collaboration between Governments and Industry

Collaboration between governments and the private sector have emerged as an important way to support both the development of a commercial sector space economy while also meeting national space goals. These partnerships can take many forms and vary greatly in terms of focus, structure, and actual outcomes.

Early examples of some public-private partnerships that focused on specific goals of space agency include:

Commercial Orbital Transportation Services (COTS): In this well-known example of a successful public-private partnership, starting in 2006, NASA worked with commercial industry partners to develop and demonstrate new commercial cargo space transportation systems to low Earth orbit. Markedly different from previous approaches to capability development,a key feature of COTS was that the commercial firms invested their own funds into the development of spacecraft and launch vehicles and owned the resulting capabilities, while NASA served as a seed-money investor and advisor during the development. For more information on the program, see the NASA-published Commercial Orbital Transportation Services document for more details and lessons learned.

Figure 2.3 | Members of the public watch the launch of a SpaceX Falcon 9, a rocket whose development included pre-arranged commitments by NASA to purchase several operational flights as part of the COTS program. *Credit: Max Alexander*

TerraSAR-X: Another example of a successful public-private partnership between the German Aerospace Center and Airbus Defence and Space is the TerraSAR-X, a satellite that produces images using a synthetic aperture radar at one-meter resolution. The government was responsible for development and operation of the ground segment as well as the instrument calibration and satellite operations after launch. Airbus assumed much of the initial risk of developing and deploying the satellite but continues to receive revenues from the data still being produced.

Other types of collaboration can involve multiple rounds or ongoing opportunities that encompass a broader variety of goals. These programs seek to encourage and leverage emerging markets and capabilities to better serve national priorities. A few ongoing examples include:

Space Technology Mission Directorate Solicitations: NASA's Space Technology Mission Directorate seeks out partnerships to encourage and support priority space technologies and capabilities. As an example, the Tipping Point solicitations focus cost-share proposals for industry-developed space technologies where an investment may significantly mature the technology, increase the likelihood of infusion into a commercial space application, or bring the technology to market for both government and commercial applications. NASA's Techport provides information on prior and current solicitations.

ESA Space Solutions: With a focus on green transition and a digital, sustainable future, ESA Space Solutions works with businesses across the European continent to develop space-enabled solutions. Projects can include kick-starts, feasibility, studies, and demonstration projects and support ranges from financial to technical expertise to networking access.

Japan Space Strategy Fund: Established in March 2024, the Space Strategy Fund is an effort across several agencies and ministries in Japan, that will focus on the following issue areas: satellites, space exploration and space transportation aimed at market expansion, solving social issues, and pioneering new frontiers. Over ten years, this fund is expected to distribute more than $6 billion to companies through public calls.

space or non-governmental private actors, is that the governmental policy toward the private or commercial space sector will have a significant impact on the business success of those private space ventures.

Property Rights

Ownership and control rights to space objects launched by a state and registered by it are protected for that state (and are protected for its nationals as long as their state extends those rights to them). These rights may also apply to a state that procures a launch from another country. Other tangible property rights are more uncertain.

Concerning rights to resources, while samples of space materials may be obtained, commercial rights to extracted natural resources in space are widely debated and, so far, untested. As described in Chapter One: Core Treaties, the Outer Space Treaty prohibits claims to sovereignty over any celestial body. Therefore, states are effectively prohibited from granting title to any real property beyond Earth. States retain jurisdiction over their nationals, however, and this means that states have the power to protect the commercial operations of their nationals from interference from others with the same nationality. Additionally, a growing number of states are recognizing the rights of private sector entities to extract and use space resources, without also giving them full property rights. This is the strategy employed by the United States in crafting its U.S. Commercial Space Launch Competitiveness Act of 2015, and also enshrined in the principles of the Artemis Accords that have been adopted by more than forty-three countries as of May 2024.

The debate over rights to material extracted from a celestial body is complicated by differences over the meaning of the ban on "appropriation" in the Outer Space Treaty. For some, that means a prohibition on assuming any property right for off-Earth material. For others, there is a clear distinction between the use of resources extracted or harvested from a celestial body, such as regolith or water, and ownership of the body itself. Since the United States, Russia, China, and Japan have all obtained material from celestial bodies, returned it to Earth, and exercised full ownership and control of it, any ban is shown by practice to not be absolute.

For the immediate future, then, it appears that property rights to material obtained from celestial bodies will largely be determined by national legislation, and that those rights will pertain only within the territorial and personal jurisdiction of the legislating state. As a cautionary note, business plans developed with the intention of exporting off-Earth material or products derived from it should ensure that the sovereigns with authority over the intended export markets will permit the sale of such material and products. No rules specific or unique to space activity exist for intellectual property. In general terms, the rules are the same as those that would apply for terrestrial activity.

PUBLIC ADMINISTRATION AND NATIONAL OVERSIGHT

As explained in the previous chapter, states bear international responsibility and liability for damage caused by the space activities of their nationals. They also are tasked with oversight and continuing supervision of their national space activities, including space activities conducted by non-governmental actors. States use national legislation and regulations to fulfill these international legal obligations, often referred to as their Article VI obligations (after the relevant article in the Outer Space Treaty). While establishing national policy rationales and objectives, as discussed previously in this chapter, is important, there is also a pressing need to create the organizational structures to implement these policies and provide oversight of day-to-day national space activities.

National Regulators

The Outer Space Treaty obligates state parties to authorize, license, and continually supervise national activities for conformity with international law, but it is at the discretion of each state's government to determine which agencies are tasked with this regulation. In some countries, these responsibilities are divided among several different agencies.

In the United States, for example, the Federal Aviation Administration (FAA) has responsibility for commercial launches, the Federal Communications Commission (FCC) deals with commercial telecommunications and frequency allocations, the National Oceanic and Atmospheric Administration (NOAA) regulates remote sensing, and the Department of State and the Department of Commerce share responsibility for export control. In the United Kingdom, the United Kingdom Space Agency is responsible for the civil space program while the Civil Aviation Authority is an independent regulator for space. Deciding which agency is covering which activity can eliminate both gaps and redundancies in the oversight regime.

Licensing

Licensing is the standard method used by a state to authorize and regulate its national, non-governmental space activities. Individual actors must comply with requirements to obtain national licenses before undertaking space activities. The types of licenses required can vary: launch licenses, frequency-use licenses, remote sensing licenses, broadcasting licenses, etc. The criteria for obtaining these licenses can include scientific, technical, environmental, safety, insurance, and financial solvency requirements, to name a few. In most cases, private sector space activities require positive confirmation (i.e., a license) before they are allowed to occur. Licenses often include mandatory reporting or other obligations on the licensee as a way for the state to discharge its treaty obligation of providing continuing supervision under Art VI of the OST. This is different from many non-space sectors, where private sector activities are often allowed by default, and only specific types of activities, for example those that are particularly risky or harmful, are required to have permission to proceed. Understanding the licensing requirements in the applicable jurisdiction/s, the licensing processes to be followed, as well as the costs and timelines typically required to obtain the required licenses is absolutely essential for any new actor contemplating space activities.

National Registries of Space Objects

In accordance with the Outer Space Treaty and the Registration Convention, states assert ownership of their space objects by placing them on their national registries. This ownership is twofold, encompassing jurisdiction and control. Jurisdiction is a legal power to create and enforce laws and to settle claims, and is held by the state. Control is an operational power analogous to command over the space object. Article VIII of the Outer Space Treaty confers these rights:

> *A State Party to the Treaty on whose registry an object launched into outer space is carried shall retain jurisdiction and control over such object, and over any personnel thereof, while in outer space or on a celestial body. Ownership of objects launched into outer space, including objects landed or constructed on a celestial body, and of their component parts, is not affected by their presence in outer space or on a celestial body or by their return to Earth. Such objects or component parts found beyond the limits of the State Party to the Treaty on whose registry they are carried shall be returned to that State Party, which shall, upon request, furnish identifying data prior to their return.*

While the Outer Space Treaty gives states the rights and the method to assert jurisdiction and control, it does not make it mandatory; the 1975 Registration Convention, in turn, requires and obligates states to establish national registries of space objects. For states party to the Registration Convention, Article II requires the establishment of a national registry, and providing notification of the establishment of such registry to the UN Secretary-General.

National registries are sometimes created through legislative acts, either as part of general space legislation or in an act specifically for the purpose of creating such a registry, as in the cases of Argentina, the Netherlands, and Italy. National registries may also be created by executive decree or within regulations by an agency granted the power to create them.

As of May 2024, 75 states were party to the 1975 Registration Convention, and 39 of them had established national registries of space objects and informed the UN of such national registries. The European Space Agency (ESA) and European Organisation for the Exploitation of Meteorological Satellites (EUMETSAT) have also established registries. Some states place the task with their national space agency, others with their federal aviation office even if they have a national space agency, as is the case with Germany.

For states that are not party to the Registration Convention, but still wish to exercise jurisdiction and control over their national space objects, establishing and maintaining a national registry of space objects is a reliable method to assert and consolidate jurisdictional powers. For non-governmental actors, due diligence and compliance with governmental oversight likely includes determining which state will have their spacecraft on its national registry and supplying the competent authority in that state with the relevant information on their spacecraft and planned activities. For states that are party to the Registration Convention, the government is responsible for ensuring that it gets the information it needs from non-governmental actors in order to maintain its registry. The information requirements for compliance with the Registration Convention, which is covered in Chapter One: Registration of Space Objects, are modest and some states follow enhanced registration practices. For example, New Zealand has gone beyond its international legal obligations by partnering with commercial SSA data providers to gain insight on the orbital status of all its national space objects. See Case Study: New Zealand Monitoring Objects Launched from its Territory for more information.

Insurance Requirements

While states are ultimately liable for damages resulting from space activities, many have taken steps to require private sector entities to purchase insurance that can cover some or all of those potential costs (essentially indemnifying the state). Along with R&D and launch costs, insurance can be one of the highest costs associated with satellite activities and, thus, is something to consider seriously when planning for a space venture. For example, Australia, Brazil, France, Japan, South Korea, the United Kingdom, and the United States all require the purchase of insurance at varying levels. Most actors, however, do not require insurance for satellites for the entirety of while they are on orbit, but rather focus primarily on the launch and early stages of their operational lifespans, as those are the portions of space activities that traditionally have had the highest risk of damages or loss.

Waivers

There are different kinds of waivers that may be used for space activities. A cross-waiver is a legal instrument between parties where each reciprocally contracts to not hold the other party liable for any damage suffered. Cross-waivers of liability are often used in the space industry and might be used between the launch provider and the operator, or between contractors and subcontractors. Waivers have the effect of making it easier to contemplate and compute the possible liability exposure a project faces.

On a regulatory level, waivers can be granted in order to relieve operators from following a regulation that evolved after their satellite was launched. This type of waiver might also be called a "variance." Alternatively, operators can apply for a waiver from obeying a regulation that they believe to be unduly onerous or to have national security consequences. Granting waivers can be used by regulators to allow an industry to innovate; however, regulators should be certain that these waivers are legitimately needed and not being requested out of convenience or to obtain an advantage against industry competitors.

National Frequency Administration and Broadcasting Regulations

The International Telecommunication Union (ITU) deals with frequency allocation and coordination at the international level, which is covered in Chapter One: International Frequency Management. National administrators determine frequency use at the domestic level, commonly through licensing and national frequency tables, and also work within the ITU to coordinate spectrum allocations for their national entities. For example, the Ministry of Communications in India is in charge of their spectrum management, and the Office of Communications (Ofcom) in the UK provides licenses for radiofrequency use. In the United States, the FCC coordinates non-federal use of frequencies, while the National Telecommunications and Information Administration (NTIA) coordinates federal spectrum use.

In addition to working on frequency issues, these administrators can reinforce other best practices. For example, in order to receive authorization from the FCC to use a frequency, U.S. commercial satellite operators must submit an orbital debris mitigation plan that is in accordance with internationally recognized debris mitigation guidelines and even more stringent U.S. guidance. Laws and regulations pertaining to broadcasting are not limited only to space-based services and can include other sectors such as cable television. It is important for any entity undertaking space-based broadcasting activities to comply with any relevant national rules regarding broadcasting generally. For example, in Canada, companies engaged in broadcasting are required to broadcast a certain amount of Canadian content. The national regulator may also impose resolution limitations on remote sensing or limitations on power emissions.

Spectrum regulation is a part of a government's responsibility for oversight. This planning function allows for spectrum allocation, which grants use of a frequency band to a specific user, dependent upon national policies, technical characteristics of the spectrum, and international agreements. This allocation process helps ensure that the spectrum is managed and used in a sustainable way while limiting the amount of harmful interference created by its use. One growing concern is not just deconflicting between spectrum usage for different space missions but also between terrestrial communications and space sector needs. Next, spectrum engineering is the regulatory function that creates technical standards for equipment whose frequencies affect or are affected by the radio spectrum. Finally, there is spectrum compliance, which involves monitoring the use of the radiofrequency spectrum to ascertain that users are complying with technical standards and frequency allocations.

Administration of Export and Technology-Transfer Controls

States implement export control measures to meet international commitments for non-proliferation regimes, to enhance regional stability, and out of national security interests. States must decide how to administer export control laws.

In order to reliably control exports, a country must establish legal authority to do so, which would correspond to six principles: comprehensive controls,

Case Study: Export Controls in the United States

The United States has three agencies with the authority to issue export control licenses: the Departments of Commerce, State, and the Treasury. Often, exporters must go to more than one agency and must ask for multiple licenses. There is interest in streamlining this process to have one single licensing agency in charge, although this would be a complicated effort and challenging to implement.

The U.S. Department of State administers perhaps the most well-known example of export control regimes, the International Traffic in Arms Regulations (ITAR), a set of U.S. government regulations that control the export and import of defense-related articles and services on the U.S. Munitions List (USML). Businesses must register their products with the State Department's Directorate of Defense Trade Controls (DDTC), and are required to apply for export licenses and approvals for hardware and related technical data on the USML before they can be can be legally exported. The process can be expensive and lengthy and can add significant burden to commercial activities, particularly for smaller firms. Failure to comply with the ITAR requirements can lead to serious fines, jail time, and other civil and criminal penalties.

In addition to the ITAR, the United States also has the Export Administration Regulations (EAR) that apply to a much broader set of commercial dual-use technologies and data that are not covered by ITAR. Administered by the Bureau of Industry and Security (BIS) within the Department of Commerce, the EAR is seen as less burdensome but still has controls in place over many U.S. space technologies and services. Technologies controlled under EAR are listed on the Commercial Control List (CCL).

Additionally, the Department of the Treasury's Office of Foreign Assets Control (OFAC) enacts financial penalties against certain actors (states, companies, or persons), depending on U.S. foreign policy goals. It keeps lists of those who are under trade restrictions, lists which are included under the BIS Consolidated Screening List.

Satellites and related technologies present a significant challenge for export control. In the early 2000s, the U.S. Congress passed legislation that placed all satellites and space-related technologies on the USML, due to concerns over transfer of space technology to China that could be used to improve ballistic missiles. The stricter controls on export of U.S. satellite technology led to foreign firms developing their own products, which were often marketed as "ITAR-free." As a result, the global market share for U.S. satellite companies dropped precipitously. A strong push from industry led to Congress passing an updated law in 2012 that gave the White House the authority to determine which specific space technologies would remain on the USML, and which technologies would be transferred to the less onerous CCL, while retaining a prohibition on export of space technologies to specific countries. In 2014, after two years of interagency and public deliberations, the Department of Commerce announced the shift of some types of satellites and space technologies to the CCL.

However, the steps to reform U.S. export controls for satellites have not satisfied all the critics. Companies now need to determine whether or not they need to apply for a license from the State Department or the Commerce Department, and the overall system has become more complex. Furthermore, commercial satellites performing above a certain standard still remain on the USML, as do any spacecraft designed for human habitation that has integrated propulsion and spacecraft designed for satellite servicing. There continues to be an on going discussion between the U.S. space industry and the U.S. government over future changes and reforms to export control. For more information, see Introduction to U.S. Export Controls for the Commercial Space Industry.

implementing directives, enforcement power and penalties, interagency coordination, international cooperation, and protection of government dissemination of sensitive business information. Next, a country needs to establish clear regulatory procedures that include a list of controlled items. Finally, the export control system should have enforcement built into it, including transparent procedures for issuing export licenses, compliance verification mechanisms, and investigation of possible illicit exports.

Congestion in Space

As states are responsible for their own space activities and those of their non-governmental entities, national policies and administration for dealing with congestion in space are important for improving space sustainability. Addressing congestion in space deals with tracking space objects and reducing collision risk between space objects (including both debris and active spacecraft). These are key objectives of space situational awareness (SSA) systems and space traffic management and coordination efforts; and are closely related to space debris mitigation and remediation efforts.

 Case Study: New Zealand's Space Regulatory and Sustainability Platform

In 2019, the New Zealand Space Agency (NZSA) and the company LeoLabs announced their intent to build the Space Regulatory and Sustainability Platform. This cloud-based platform uses data from LeoLabs' radars in order to track satellites in low Earth orbit, as NZSA wanted to make sure that satellites launched from its territory were following New Zealand's licensing rules regarding debris mitigation; additionally, NZSA aimed to see if conjunctions were possibly going to happen in the hopes of preventing collisions that could create large amounts of orbital debris. This was part of how New Zealand interpreted its Article VI obligations from the Outer Space Treaty to provide authorization and continuing supervision of national activities in space by non-governmental actors.

The Space Regulatory and Sustainability Platform was unveiled one year after New Zealand officially became a launching state when Rocket Labs began launching from its territory in 2018. New Zealand officials wanted to make sure that any satellites it was responsible for as a launching state were complying with regulations as part of ensuring the responsible and sustainable use of space.

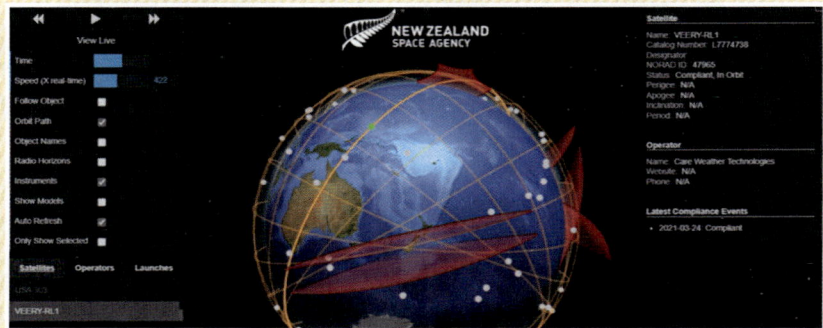

Figure 2.4 | Sample from LeoLabs and NZSA's platform. *Credit: LeoLabs*

A topic many states struggle with is potential exceptions to space debris mitigation guidelines at the national level. It may be necessary to exempt some long-running programs from specific aspects of the guidelines because portions of the programs were designed and implemented before the guidelines were adopted. States may also be inclined to exempt some new programs over concerns that implementing the guidelines will lead to increased costs or operational challenges. However, widespread exemptions would have a deleterious effect on adherence to the guidelines, which would ultimately negatively impact all space actors. If states are to make exemptions to the guidelines, they should do so through a well-defined, rigorous process that includes high-level decision-makers and clearly outlines the costs and benefits of the exemptions.

> All states are encouraged to put in place national mechanisms to implement the UN COPUOS and IADC debris mitigation guidelines for both governmental and non-governmental actors. How they are implemented can vary depending on a state's specific governmental structure.

In addition to limiting the creation of new debris, several states have also put in place policies and administrative practices to minimize the impact that existing space debris have on space activities. The United States, Russia, China, France, Germany, India, and Japan are among the states that have governmental organizations tasked with monitoring the population of space objects and predicting potential close approaches. In some cases, these organizations do so for their own governmental satellites, while in others they do so for non-governmental or foreign satellites as well. In either case, they have put in place procedures and data-sharing mechanisms for notifying satellite operators and assisting them in assessing the risk of collision and implementing any avoidance measures.

These practices are often included in the larger discussion about space traffic management (STM), but at present there is no standard national practice for implementing STM in a comprehensive manner. It is currently up to each satellite operator to determine their own tolerance for risk and to use that as a basis for determining whether to take steps to avoid a close approach with another space object. Current techniques for predicting close approaches and possible collisions in orbit are not sophisticated enough to enable mandatory maneuver policies, with the specific exception of activities such as human spaceflight.

Several states have also put in place policies and organizations for providing a national space situational awareness (SSA) capability. Developing the capability to track all space objects requires a considerable network of tracking-station locations around the world. Thus, SSA has traditionally been a government function and is often either based on existing national military or dual-use capabilities. This can create challenges for states that do not have a prior working relationship between their national security community and their civil space community, or for states that try to develop SSA capabilities as a purely civil function. Third party services and commercial companies are now operating SSA sensor networks and providing SSA information to satellite operators. Commercial SSA is further discussed in Chapter 3.

All states are encouraged to put in place national mechanisms to implement the UN COPUOS and IADC debris mitigation guidelines for both governmental and non-governmental actors. How they are implemented can vary depending on a state's specific governmental structure. Usually, implementation includes policy directives for federal agencies, a regulatory component in national law, and licensing requirements for non-governmental entities.

IN-DEPTH ANALYSIS: REMOTE SENSING POLICY AND ADMINISTRATION

Remote sensing satellites have continually sensed Earth for more than six decades, yielding a valuable repository of data about the planet which has applications in areas as far-reaching as health, climatology, and urban planning. Given its strong linkages to socioeconomic development, space-based remote sensing is a key area of activity for new and established space actors alike. In light of this, remote sensing is a useful case study highlighting the interaction between public policy and public administration and illustrates some of the approaches different countries have taken to managing this kind of activity. Additionally, new trends in remote sensing activities, especially by non-governmental actors, illustrate larger policy transformations that are useful for new space actors to consider.

Figure 2.5 | This composite image uses shortwave infrared data overlaid on a natural-color mosaic image based on Landsat 8 observations for added geographic detail of the 2023 wildfire in Lahaina, Hawaii. *Credit: NASA*

Remote Sensing Policy

Consistent with the main elements of public policy described in the beginning of this chapter, remote sensing policies primarily seek to:

- identify objectives and priorities guiding the acquisition of data about the planet;

- define roles and responsibilities of government remote sensing activities as well as related oversight obligations;

- set requirements by designating procedures private operators must follow to operate remote sensing systems; and

- identify data policies to govern the conditions of access and distribution of the data acquired through the operation of these systems.

Remote sensing policies may be included within national-level space policies, or may have dedicated policies of their own. In some cases, a government will lay out specific goals with respect to the information being collected or priority application areas, identifying departments or agencies that are responsible for acquiring research or operational datasets. Specific government agencies may also be tasked with operating specific systems. Sector guidance in the 2020 U.S. National Space Policy, for example, dictates that the U.S. Geological Survey (USGS) and National Aeronautics and Space Administration (NASA) shall cooperate to maintain an operational land remote sensing program. The policy also describes tasks related to the acquisition, archiving, and distribution of the global land remote sensing dataset.

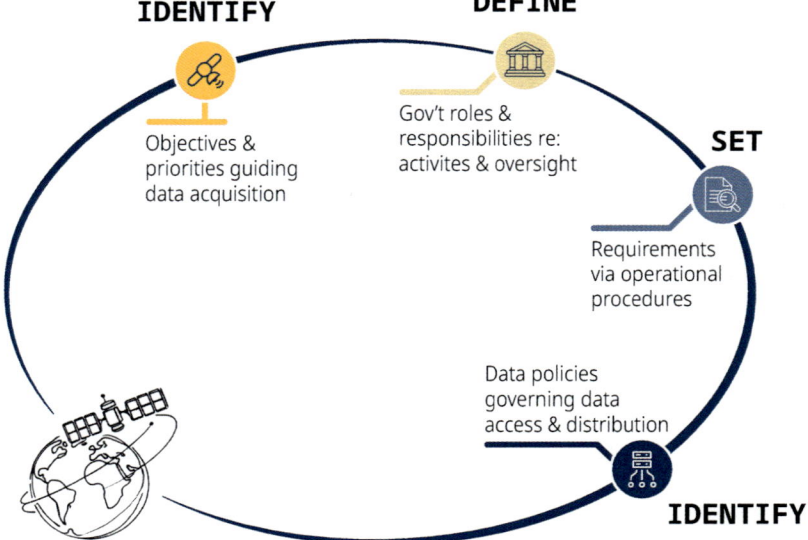

IDENTIFY

Objectives & priorities guiding data acquisition

DEFINE

Gov't roles & responsibilities re: activites & oversight

SET

Requirements via operational procedures

Data policies governing data access & distribution

IDENTIFY

Figure 2.6 | Aims of remote sensing policies.

In light of technological advances driving the proliferation of these systems and the international liability and responsibilities discussed in Chapter One, remote sensing policies also define roles with respect to the oversight of non-governmental remote sensing activities, identifying the specific department or agency and the tasks they perform in the process of authorizing and supervising a given activity. These guidelines may be further detailed in related regulations, laws, or agency-level policies. In the United States, the licensing authority for private remote sensing space systems lies with the Secretary of Commerce, a task that has been delegated to the NOAA for implementation, and whose principles are captured in the National Commercial Remote Sensing Policy. NOAA agency-level policies specify the principles guiding related activities, such as a partnership policy among government, academia, and the private sector in the provision of environmental information and services.

Policies also allude to coordination processes necessary to orchestrate the different elements involved in managing remote sensing activities, which, in addition to those common for any satellite mission (research, development, launch, operations, etc.), include tasks specific to the processing, archiving, and distribution of Earth observation data. Institutional coordination is particularly necessary in the field of remote sensing because of the diversity of users and stakeholders who routinely derive valuable intelligence from this information. Understanding needs across these different user communities is an often challenging but crucial task that feeds into this coordination process to improve the value of investments in remote sensing programs.

Oversight of Non-Government Remote Sensing Activities

Rapid technological advances often drive the evolution of remote sensing policies, particularly given the growth of high-resolution imaging satellites from non-governmental sources. Remote sensing policies primarily seek to advance national (including commercial) remote sensing activities for the provision of services, imagery/data, or value-added products while balancing national security and foreign policy interests. To do this, the policy will specify procedures that non-governmental operators must meet to be allowed to operate space remote sensing systems, and the limitations on such activities. Access to the data acquired by these systems, whether freely or commercially, is also subject to specific limitations imposed by the oversight authority.

Canada's Remote Sensing Space Systems Act of 2005, for example, details the procedure by which an operational license may be canceled or temporarily revoked when it is determined to be "injurious to national security, the defence of Canada, the safety of Canadian Forces or Canada's conduct of international relations" or "inconsistent with Canada's international obligations." In some countries, a license may not be revoked but the operator may be required to temporarily cease operations during crisis or conflict—sometimes called "shutter control"—or to refrain from sensing or distributing data on areas of the world deemed to be sensitive by the licensing authority. As more commercial companies provide Earth observation data, it is increasingly becoming a challenge to limit the dispersal of these images and thus attempts to do so via regulation may concurrently become increasingly rare. Another consideration is whether the data that a state is attempting to restrain is available internationally; if so, domestic restraints will provide little effect other than harm to the domestic industry.

Permissions are often granted through licenses issued after the operator has committed to meeting certain operational and even disposal procedures, and sometimes following interagency review. As an example, a U.S. commercial remote sensing license application available on the NOAA Commercial Remote Sensing Regulatory Affairs website highlights the basic requirements, including the following:

- **Corporate information:** contact information and other details about the business, a description of significant agreements with foreign nations or persons, etc.

- **Launch information:** proposed launch schedule, anticipated operational date, orbital parameters, etc.

- **Space segment information:** anticipated resolution and swath width of sensors, onboard storage capacity, anticipated system lifetime, etc.

- **Ground segment information:** proposed system data collection and processing capabilities, transmission frequencies, plans for protection of uplink and downlink, etc.

- **Other information** including financial information about proposed commercial data distribution policies, a plan for post-mission satellite disposal, etc.

Data Policies

Data policies are a key component of remote sensing policies since these specify the access and distribution rights and obligations of data acquired through these activities. Generally, policies make most government-acquired remote sensing datasets available for scientific, social, and economic benefit by making the data available to users across government, academia, and the private sector. The European Union's Copernicus system's data is available to users "on a full, free, and open basis," and is referred to on its website as a "public good." Bilateral and multilateral data exchange programs also exist to facilitate the sharing of specific datasets among operators and users of partner countries or when such exchange helps address shared challenges.

The World Meteorological Organization (WMO), for example, facilitates the international exchange of meteorological and related data and products, including those derived from space-based systems, as tied to "matters relating to safety and security of society, economic welfare and the protection of the environment." National open-data policies may also apply, usually as part of a larger policy governing the access and use of government-funded data that is not limited to space, and may include data acquired through airborne or in-situ platforms.

Even with the proliferation of open data access policies, remote sensing policies include language specifying conditions of restriction on access or redistribution of datasets, particularly driven by national security concerns. The main driver for these different policy elements is the inherent dual-use nature of remote sensing technologies, which enable applications across civil, commercial, and military domains. Some satellites or systems serve the needs of both civil and military users, thus making them "dual use." However, even while a satellite or a system may be designed to serve exclusively civil needs, the dual-use nature of the technology remains, as the data gathered can be aggregated or reused to feed into applications for military purposes. Consequently, and to balance national security concerns associated with access to potentially sensitive information, data policies typically specify resolution or temporal restrictions for the distribution of high-resolution data or imagery, including those from commercial operators and providers.

The exchange or redistribution of these datasets may be subject to additional requirements and examined on a case-by-case basis. In Germany, Earth observation data acquired through "high-grade" systems are subject to the German National Data Security Policy for Space-Based Earth Remote Sensing Systems, and distribution is allowed depending on the level of "sensitivity" of the data. India requires authorization from IN-SPACe for the dissemination of any remote sensing or Earth observation data of ≤30 cm Ground Sampling Distance (GSD) at nadir.

Broader Policy Context

Driven in large part by technological advances, data policies—and their application through licenses and other legal mechanisms—will remain a focal point in the evolution of remote sensing practice currently manifested through the emergence of multiple sources of non-government data and services. The expansion of non-government actors in the full value chain of remote sensing activity—from research and operations to data processing and archiving—is one of the trends raising new policy and regulatory questions. Another important trend is the proliferation of geospatial products and services that result from the aggregation of multiple datasets, which may come from several data providers and are often collected from various space-based, airborne, and in-situ platforms. In a context where space activities represent a portion of the remote sensing activities that governments must oversee, space-derived data and services may be subject to regulation or oversight by multiple government agencies and be governed under different legal regimes. For example, privacy debates arising from uninhabited aircraft systems (UAS) in the United States are beginning to expand to include discussion of similar concerns related to small satellites, despite the fact that these systems currently operate in distinctly different legal domains. These and other developments suggest that in some countries, new rules may emerge that apply to specific types of remote sensing applications or the kinds of data being collected, rather than specific collection platforms. In this context, new space actors should be aware of this broader policy context and should pay attention to how public administration practices—encompassing policy, legal, and regulatory measures—from non-space domains may apply to this key area of space activities.

CHAPTER TWO ADDITIONAL RESOURCES

See below for additional resources, links, and documents referenced in Chapter Two.

Convention on Registration of Objects Launched into Outer Space: https://www.unoosa.org/oosa/en/ourwork/spacelaw/treaties/introregistration-convention.html

DLR TerraSAR-X: https://www.dlr.de/en/research-and-transfer/projects-and-missions/terrasar-x

ESA Space Solutions: https://business.esa.int/about-us

European Organisation for the Exploitation of Meteorological Satellites (EUMETSAT): https://www.eumetsat.int/

Federal Aviation Administration (FAA): https://www.faa.gov/

Federal Communications Commission (FCC): https://www.fcc.gov/

German Space Situational Awareness Center: https://event.dlr.de/en/ila2022/weltraumlagezentrum/

Handbook of Space Capability Development: https://www.cgdev.org/sites/default/files/handbook-space-capability-development.pdf

Indian National Space Promotion Authorization Center (IN-SPACe): https://www.inspace.gov.in/inspace?id=inspace_index

Indian Space Policy 2023: https://www.isro.gov.in/media_isro/pdf/IndianSpacePolicy2023.pdf

Indian Space Research Organisation: https://www.isro.gov.in/

Japan Space Strategy Fund: https://fund.jaxa.jp/about/

NASA Artemis Partnership Announcement: https://www.nasa.gov/news-release/nasa-announces-partners-to-advance-tipping-point-technologies-for-the-moon-mars/

NASA Earth Data – Remote Sensing: https://www.earthdata.nasa.gov/learn/backgrounders/remote-sensing

NASA NASA Commercial Orbital Transportation Services (COTS): https://www.nasa.gov/wp-content/uploads/2016/08/sp-2014-617.pdf?emrc=81adc8

NASA NASA Tipping Point Solicitations: https://www.nasa.gov/space-tech-industry-partnerships/

National Commission on Space Activities of Argentina: https://www.argentina.gob.ar/ciencia/conae

National Oceanic and Atmospheric Administration (NOAA): https://www.noaa.gov/

New Zealand Space Agency: https://www.mbie.govt.nz/science-and-technology/space

NOAA Commercial Remote Sensing Regulatory Affairs: https://www.space.commerce.gov/regulations/commercial-remote-sensing-regulatory-affairs/licensing/

Remote Sensing Space Systems Act of 2005: https://laws-lois.justice.gc.ca/eng/acts/r-5.4/index.html

The Asia-Pacific Regional Space Agency Forum (APRSAF): https://www.aprsaf.org/

U.S. Department of Commerce – Introduction to U.S. Export Controls for the Commercial Space Industry: https://www.space.commerce.gov/wp-content/uploads/2017-export-controls-guidebook.pdf

U.S. National Remote Sensing Policy: https://www.space.commerce.gov/policy/national-space-policy/

UK Civil Aviation Authority: https://www.caa.co.uk/

UK Space Agency: https://www.gov.uk/government/organisations/uk-space-agency

United Arab Emirates Space Agency: https://space.gov.ae/en

NATIONAL SPACE POLICY & ADMINISTRATION

CHAPTER THREE

Responsible Operations in Space

CHAPTER THREE FOCUS

The preceding chapters discussed the international legal framework within which national activities exist, and how states establish space policy, perform interagency coordination, and supervise and regulate their national activities through legislation, licensing, and authorization.

This chapter focuses on space activities themselves, and is divided into pre-launch, launch, in space, and end-of-life issues. As a result, it is also more technological and operational in perspective than previous chapters. It gives concrete guidance to new actors in space—whether they are new states, start-up companies, or academic and university-led projects—as they begin their space activities. The best practices contained in this chapter are the types of behaviors that responsible actors will implement if they want to conduct successful space operations while also preserving order, fostering cooperation, and ensuring the long-term sustainability of space activities.

There are multiple sets of best practices and standards developed by the private sector that apply to responsible operations in space, with more developed or refined all the time. Expanding commercial activities may not fit clearly within existing regulatory frameworks or processes, placing additional importance on responsible operations practices and the need for national regulatory and policy attention to private sector space activities.

Walt Everetts

RETIRED VICE PRESIDENT
Space and Ground Services at Iridium

INTRODUCTION

In 2009, an abandoned, uncontrolled Russian communications satellite collided with one of Iridium's active communications satellites in the first ever on-orbit collision between satellites. The collision destroyed both satellites and produced thousands of pieces of debris which continue to pose a risk to other satellites in low Earth orbit. I was there to see it, and I am proud of how my company led the industry in taking steps to ensure such an accident does not happen again.

The unfortunate incident was a wake-up call for the space industry. With nearly 10,000 active satellites in various Earth orbits, today's space environment is significantly more active and complex than it was in 2009. Increasingly, the digital and informational appetites of our critical infrastructure and of the many products and services we use each day (and often take for granted) are dependent on the uninterrupted operation of satellites.

The space environment, and the services we derive from it, requires close coordination and timely information sharing to improve operational safety and reduce risk. I am grateful that the space community has significantly advanced the development and adoption of responsible and sustainable operations approaches, and has allocated the resources, including space situational awareness data, necessary to support those approaches, but more needs to be done.

This chapter describes the core behaviors and approaches that are necessary to conduct responsible space operations while ensuring the long-term sustainability of the space environment. Current and future space actors have a responsibility to protect the continued accessibility and stability of the space environment. All phases of a space mission, including launch, on-orbit operations, ground segment activities, and end-of-life protocols require sustainable practices such as enhanced space debris mitigation, operator transparency and information sharing, and design for demise.

Regulations, policies, guidelines, and best practices all have a place in enabling sustainable use of space. The effectiveness of these measures depends upon all space actors taking accountability for their space assets in recognition of the shared nature of space.

CHAPTER THREE
Responsible Operations in Space

PRE-LAUNCH

Space activities begin well before a satellite is actually launched into space. In addition to designing and building the spacecraft, there are a number of policy, legal, and administrative steps that need to be taken into consideration. The following topics are deeply connected to both the design of satellites and operations in space and should be considered well before launch and the commencement of space operations.

Mission Architecture Design and Planning

Many choices that are made during the design of a satellite or a constellation will have significant impacts on the space environment. While satellite power systems, propulsion, guidance, and thermal management are widely discussed in many satellite engineering texts and handbooks, their impact on space sustainability and responsible space operations are not always emphasized.

From the very beginning, operators need to develop a plan of how to manage spacecraft operations at the end of service. International standards exist for how long a satellite and any associated objects should remain on orbit post-mission, and a growing number of governments are including a disposal plan in their licensing requirements. Design choices such as whether or not a satellite has propulsion, its operational altitude, and whether or not it is designed to be serviced all impact the options available for disposal.

Orbital carrying capacity is also a topic of growing concern. While space itself is vast, many satellites are being placed into the same or similar orbits, which may also include constellations of hundreds or thousands of satellites. Even if they do not result in a collision, the increasing number of possible

> Many choices that are made during the design of a satellite or a constellation will have significant impacts on the space environment. While satellite power systems, propulsion, guidance, and thermal management are widely discussed in many satellite engineering texts and handbooks, their impact on space sustainability and responsible space operations are not always emphasized.

conjunctions among these satellites can create added costs and operational complexity for the operators involved. While there are no agreed-upon limits of what the carrying capacity of various orbits are, the congestion of the "local neighborhood" needs to be taken into account when selecting a mission altitude.

Another important consideration is the impact a satellite or constellation will have on the Earth's environment, both on the way up and on the way down. Ongoing research suggests that the increasing number of rocket launches and re-entering satellites may have negative impacts on the Earth's atmosphere. However, those findings are still inconclusive and at this time not suitable for drawing any specific best practices or recommendations.

While there are a growing number of space actors planning for cislunar and lunar space activities, standards and best practices for operating sustainably in those regions have not yet been developed. Space activities in cislunar and lunar regions operate under a very different set of physical and environmental characteristics than in Earth orbit, and it is likely a different set of sustainable practices will need to be developed and implemented. To some extent, it will only be feasible to develop such standards and best practices based on the accumulated experience of those conducting cislunar and lunar surface operations. However, while such experience is being accumulated, the actors involved should apply a general precautionary principle of conducting their operations in a manner that should not cause harm to the environment or harmful interference with the operations of other actors.

Developing a Post-Mission Disposal Plan

Spacecraft operators have multiple options for how to dispose of their satellite at end of life. The most common method, known as direct re-entry, is to actively maneuver it into the Earth's atmosphere on a controlled trajectory. However, this method requires significant propulsion capabilities and fuel reserves, which can be expensive and is not suitable for all missions.

Another common method is to use natural decay into the Earth's atmosphere on a reasonable timeline. This is normally done with a smaller disposal maneuver than direct re-entry, or by having the satellite operate at an altitude which is already compliant. While on orbit, the satellite may pose a collision risk to other space objects and the actual time it takes to re-enter can be hard to determine as a result of variations in solar activity.

If direct or natural atmospheric re-entry is not possible, another option is to place the satellite in one of the designated storage orbits. These orbits are designed to reduce the likelihood that the object will pose a risk to operational satellites in active regions. The most well-known of these regions is the so-called "geosynchronous graveyard" located roughly 300 kilometers above the active GEO belt. If this method is chosen, it is critical that the satellite be entirely passivated through the elimination of all potential energy sources, including batteries and remaining fuel, to reduce the likelihood of explosions that can create debris in the future. Options for active debris removal—use of an external service provider to remove a spacecraft from orbit—are emerging, and may provide backup or last-resort options should other methods of post-mission disposal be not viable. To facilitate future active debris removal, spacecraft can be designed for removal, in which features such as grappling points or fiducial markers are installed to aid in potential removal operations.

Design for Demise

Design for Demise is a method of satellite design with the goal of ensuring each component of a satellite will be completely destroyed during the heat of re-entry. By designing for demise, satellite operators can avoid having to conduct a controlled re-entry, which can lengthen the mission lifespan, lower the cost of development, and reduce the mission ground-support costs. Design for Demise is a good approach for ensuring compliance with the 1-in-10,000 casualty risk from space debris re-entry, which is commonly accepted as a threshold by space agencies. The International Organization for Standardization (ISO) has a standard (ISO 27875:2019) that can be applied at the planning, design, and review stages of satellite development to assess, reduce, and control the potential risk that spacecraft and launch-vehicle orbital stages pose during re-entry.

Cyber Security

Satellites and their ground control systems involve increasingly complex and powerful computers and thus are susceptible to cyber attacks. Historically many satellites used custom hardware and software and dedicated communications links that made it harder for cyber attackers to get in. Today, many space systems are using standard commercial hardware and software and may be connected to terrestrial communications networks, all of which can increase the attack surface a hostile actor could exploit.

Cyber attacks against space systems comprise several different categories. The first is a supply chain attack that inserts faulty, counterfeit, or malicious microelectronics and materials into the manufacturing process, which can result in vulnerabilities or flaws in the final system that an attacker can leverage. Another category is a cyber attack directed against the links between satellites and ground control stations, such as gaining control of the software used by a commercial antenna network. A third category is an attack against terrestrial computer systems that are used to process, analyze, and disseminate space data, such as networks used to distribute space-based weather imagery. Finally, cyber attacks can be directed against the end user segment, bypassing both the space and ground control segments. For example, malicious software installed on a smartphone can disrupt that device's ability to use satellite navigation signals accurately.

To date, there have been few publicly confirmed cyber attacks against space systems, and many satellite operators are reluctant to talk openly about vulnerabilities or lessons learned from past attacks. This makes it difficult to provide specific recommendations for steps that can be taken to harden space systems against cyber attacks. That said, there are some noteworthy resources that should be considered. The Secure World Foundation's Global Counterspace Capabilities report assesses the current and near-term future capabilities for each country, along with their potential military utility. One is the MITRE ATT&CK® knowledge base, which is a publicly accessible repository of cyber attack tactics and techniques that can aid in developing a threat model for a specific space system. Another useful resource is NASA's Space Security Best Practices Guide, which provides cyber security guidance for space systems, missions, and projects of all sizes.

Pre-Launch Payload Testing

Launching a satellite into space exposes it to significant vibration and acoustic forces, shock, coupled loads, and thermal and electromagnetic effects. Satellite designers and engineers need to reference a launch vehicle's user guide for information about the environment during launch, and properly test a spacecraft to make sure it will survive the launch. These risks may also extend into the early phases of a satellite's space activities, particularly if it will be undergoing weeks of maneuvering to reach its final orbit. Steps can be taken during the design, engineering, and testing phases of satellite development to prepare a spacecraft for successful deployment.

During the design phase, it may be advisable to select a satellite bus—the main body of the satellite—that has significant legacy space experience. Commonly used satellite designs should have significant data collected about how the spacecraft structure and components handle a launch environment. Furthermore, using a proven satellite and launch-vehicle combination further reduces the risk of payload deployment failures.

Spacecraft must be designed to handle the vibration and acoustic effects generated by rocket motors as a satellite is launched into space. A spacecraft will be exposed to at least three types of vibro-acoustic environments that occur during launch, including random vibration, sine vibration, and acoustically induced vibration. The greatest vibro-acoustic effects are present during the first minutes of a launch, when pressure and reverberations are the strongest. This is followed by flow noise as air streams over the payload fairing, causing reverberating sound within, and is particularly strong during flight through high-dynamic pressure, such as transition through the sound barrier. Information about the vibro-acoustic environment of a launch system can be found in a launch vehicle's user guide.

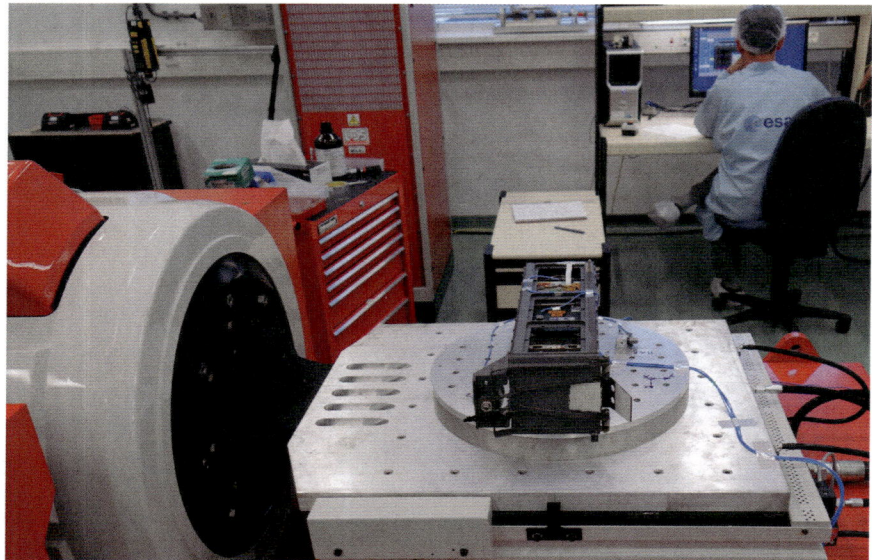

Figure 3.1 | An example of a vibration testing station. *Credit: ESA*

Most ground testing regimes simplify the launch environment and test to the most extreme conditions, not the specific mission profile. Therefore, if a spacecraft design is susceptible to vibrational effects, a non-standard, more spacecraft-specific vibration testing regime should be developed. Vibration effects can be mitigated during the design and engineering phases by incorporating motion control solutions to aid in attenuating sine vibration events and random vibration created by the launch vehicle.

Spacecraft will experience short, intense transient accelerations with broad frequency content and a very short duration, generally less than 20 milliseconds. These shocks occur during specific flight actions, such as the separation from a spent stage with an explosive charge, and can be straightforwardly modeled and tested on the ground. The hazards of shock can be mitigated by using non-pyrotechnic bolt-cutter-type release mechanisms.

In addition to taking account of the effects of vibro-acoustics and shock generated by a launch vehicle, it is also necessary to understand the coupled loads generated by the interaction of a launch vehicle and spacecraft as a complete structural system. There are a variety of methods to model coupled loads, but their quality and accuracy are highly dependent on the spacecraft's structural dynamic model and data gathered from previous launches. During the course of a satellite's design and launch-vehicle selection process, it is wise to iteratively update a coupled load's model as the spacecraft design matures and more data about the force environment of a launch system are collected.

During the launch and orbit raising phase, the thermal environment has to be maintained within the bounds for which the electronics and deployment mechanisms have been designed and qualified. Different methods are used to ensure this. At the launch pad, the capsule of the launch vehicle is air-conditioned or heated to maintain the limits of the temperature excursion. After the payload fairing is jettisoned, the launch vehicle rotates to expose the satellite to the sun to keep the temperature inside of the satellite within the allowable temperature range acceptable to the electronics, and to warm the deployment mechanisms.

During launch, spacecraft will be exposed to various electromagnetic environments, including energy from tracking radars, launch vehicle radio-frequency (RF) transmitters, flight through regions of energetic protons, and atmospheric lightning. Therefore, during the engineering phase, it is important to strictly adhere to electromagnetic design specifications, and to model possible occurrences of electrical interference. System-level compatibility between a spacecraft and launch vehicle is addressed through integrated avionics testing during manufacturing, with attention to electrical bonding and isolation requirements for a launch vehicle. Full system integration testing occurs at the launch site.

The Links Between Testing and Anomaly Mitigation

The importance of the design, manufacturing, and testing of a spacecraft cannot be overemphasized when it comes to mitigating operational anomalies in space. For all but human missions, these phases present the only opportunity for true "hands-on" and re-engineering time with the system.

The following list provides best practices to consider in developing a process from the pre-operational phase to the phases for reducing occurrences of, and impact from, certain space anomalies:

☑ *Perform a detailed Failure Modes and Effects Analysis (FMEA) at multiple phases of design and eliminate single-point failures wherever possible.*

☑ *Leverage FMEA results to develop robust and detailed operational procedures and execute these during the integration and test (I&T) phase to characterize system behavior with an opportunity to update prior to launch.*

☑ *Catalog and save all documentation and test data including vendor-provided material. This information can be critical to determining the root cause of a failure.*

☑ *Develop a flight-like simulator and/or engineering model of the system. A robust simulator is an invaluable tool for testing complex operational procedures, validating firmware and software upgrades, and performing detailed root cause investigations.*

☑ *Ensure the design of the spacecraft provides ample data for diagnosing anomalies by incorporating sufficient telemetry access points providing insight from every unit onboard a vehicle and developing detailed and well-organized telemetry formats.*

Practices such as these help the satellite operator understand the risks inherent in the mission profile during all phases of a mission and design to mitigate those risks.

Space Mission Licensing

A non-governmental satellite operator or other responsible entity needs to obtain one or more licenses for their space activities. These licenses may include (and are not limited to) radiofrequency, remote sensing, payload authorization, and launch activities. National governments generally administer this access through licenses that satellite operators are required to obtain before they are permitted to launch their system/s. Launch operators must also obtain licenses, which might separately pertain to launch activities and to re-entry activities. See Chapter Two: Licensing for more information.

LICENSING REQUIREMENTS

Licensing requirements affect most aspects of space operations, including telecommunications and remote sensing operations, launch services, and the operations of satellite ground stations on Earth (satellite Earth stations). The issuing of licenses is one of the means states use to maintain compliance with their treaty obligations, as discussed in Chapter One: International State Responsibility. Licenses cover a range of topics including spectrum access, national security oversight, compliance with insurance and safety requirements, and space debris mitigation guidelines, as detailed in Chapter Two: Licensing. Satellite and launch operators are responsible for applying for and securing licenses from the relevant national regulatory authorities where they are headquartered or where they will be conducting operations. The regulatory authorities responsible for issuing licenses vary by country and domain of operations, and may include government departments or statutory bodies, national space agencies, national telecommunications agencies, and national trade or economic agencies.

FREQUENCY LICENSING

In the satellite telecommunications segment, a primary purpose of licensing requirements is the coordination and allocation of radiofrequency spectrum on a domestic and international basis. Operators seeking to deploy a satellite communications system must apply for a license to operate that system. As spectrum is a limited resource, the licensing process acts to ensure fair access to that resource while providing a mechanism to limit the potential for interference between satellite systems, and between satellite systems and terrestrial uses of the same or adjacent radio frequencies.

As covered in Chapter Two: Public Oversight and National Administration, regulators have generally implemented a licensing regime that ensures coordination and compliance with International Telecommunication Union (ITU) policies and regulations. In many jurisdictions, the regulator responsible for issuing a license to a communications satellite operator is also the authority responsible for making ITU submissions for that country. This is not, however, always the case; for example, in the United Kingdom (UK), the Office of Communications (Ofcom) is responsible for ITU filings, while the Civil Aviation Authority (CAA) is the licensing authority.

In general, any operator seeking to operate a satellite system that will receive or transmit data (including command and control linkages) over the radio-frequency spectrum must apply for a license from its relevant regulatory agency. Typically national administrations also require market access licenses for communications services. This means that service providers wishing to offer telecommunications services in a particular jurisdiction must apply to that country's regulator for permission to offer those services, even if their actual space operations are licensed elsewhere. Countries may impose space operations or debris requirements as part of the market access licensing process. In all cases prospective operators must provide a range of technical and business information to the regulator when submitting license applications.

In most cases, license applications must contain technical data describing the system, including the spectral bands to be used, a planned implementation timeline, and information concerning financial ability to construct, launch, and operate the system. Applications may also require detail on the steps to be taken to reduce the potential for harmful interference through coordination with other operators, as well as a post-mission disposal plan that takes into account space debris mitigation guidelines.

Some regulators also require operators to obtain licenses for the ground stations used to communicate with the satellites, including end-user terminal equipment (sometimes referred to as "Earth stations"). Earth station licenses serve to reduce the potential for radiofrequency interference, in particular interference with other terrestrial applications, and may also include provisions to evaluate interference with other applications, such as aviation. Earth station license applications typically require similar technical and business details to satellite network applications. For end-user terminals, the licensing authority may issue blanket licenses covering technically identical equipment.

Figure 3.2 | Example of a satellite ground station that would require licensing to communicate with satellites above. *Credit: NASA JPL*

Satellite operators, especially small companies or organizations operating single or small numbers of satellites, may choose to procure or purchase their ground stations as a service. A number of companies and service providers offer ground segment operations and infrastructure as a commercial service, in which capacity on common ground stations and antennae networks are offered on a shared basis. Satellite operators may procure time or services from these networks for both tracking telemetry and control capabilities and for data downlink and uplink. Such services might offer cost or efficiency savings to operators, as compared to building their own infrastructure. In some cases these ground-segment-as-a-service offerings might include regulatory support to assist in obtaining required spectrum licenses. Operators using this model should ensure that they understand which licensing requirements pertain to their operations.

Another option is to route a satellite's communications through a space-based service instead of a dedicated ground system. In this model, the satellite communicates with another satellite, or constellation of satellites, which in turn relays the information received to their own ground stations. Similar to ground stations as a service, one advantage of this model is it might help reduce some of the space-to-ground licensing requirements, although the space-to-space communications may still require licensing.

REMOTE SENSING LICENSES

In compliance with the international regime discussed in Chapter One and also discussed in the in-depth analysis on Remote Sensing at the end of Chapter Two, national governments may also require commercial remote sensing satellite operators to apply for a license covering the imaging capabilities of the satellite system. These licenses may be issued by authorities separate from those responsible for the communications systems aspects. Remote sensing licenses are typically required to ensure coordination with national security policies. Required information to be submitted may include system technical details; expected dates of operation; launch information; data acquisition, access, and distribution plans; data pricing policy; planned agreements with foreign entities; and a post-mission disposal plan. Remote sensing licenses may apply conditions on the operator, such as resolution restrictions and the ability to restrict imaging of designated national territory.

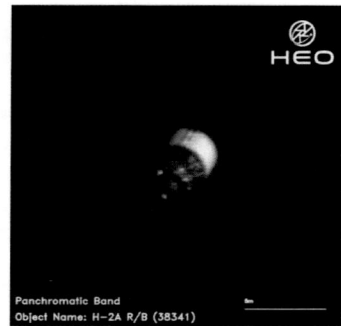

A relatively new aspect of remote sensing licenses is non-Earth imaging (NEI)—using a camera on a satellite to image other space objects. While the ability to do NEI, commonly referred to as satellite-to-satellite or "sat-squared" imaging, has existed since the early days of space activities, it was often considered a classified national security capability and commercial satellites were often prohibited from doing it. However, the proliferation of global commercial remote sensing systems has made

Figure 3.3 | Image of an H-2A rocket body taken by a space-based sensor. *Credit: HEO Robotics*

these restrictions increasingly difficult to enforce, and there is growing acceptance of NEI as a commercial product. That said, some jurisdictions still require a remote sensing license before authorizing commercial NEI services.

LAUNCH AND RE-ENTRY LICENSES

Entities providing commercial launch services are typically required to obtain a launch license from a national authority, which may differ from the authority responsible for other space-related licenses. Launch licenses may be specific to launch operations or re-entry operations, and may have varying requirements based on whether the launch (or re-entry) vehicle is experimental or operational, and whether it is expendable or reusable.

Launch and re-entry licenses authorize an operator to conduct one or more launches or re-entries defined by a specific set of operational parameters, which are codified in (and authorized by) the license. These parameters generally include, but are not limited to: mission names, intended launch windows and trajectory, parameters for the payload/s intended and final orbits, ground and flight safety plans, accident investigation plans, and re-entry windows and trajectory (if applicable). Typically, operators are also required to submit information demonstrating that their intended launch operations are in compliance with environmental policies, export control regulations, other licensing requirements (e.g., frequency and remote sensing), and insurance and liability coverage obligations.

In order to obtain a license, the launch providers may request information from the operators of the satellites to be launched. The process of obtaining a launch license entails multiple steps and submissions to the applicable regulatory authority. Such authorities often offer pre-application consultation services so that operators are aware of the steps and information required before they initiate the process.

Launch and re-entry licenses serve numerous purposes. They act to protect public safety interests, including protection of third-party safety on the ground and coordination with air traffic management functions. The licensing process provides national authorities with the ability to review the intended launch operation against national security considerations and other national regulations and requirements. The launch licensing process also acts to ensure that national authorities collect the information necessary to satisfy applicable international obligations and reporting requirements.

THE LICENSING PROCESS: OBTAINING THE NECESSARY LICENSE/S

Depending on the nature of a given space activity, one or more licenses may be required. In some cases, the issuing of a given license may be contingent on another license being issued first, so sequencing the licensing applications may be required, and more than one jurisdiction may be involved. For example, a launch service provider operating under one jurisdiction may be required to furnish licenses pertaining to the payloads to be launched, which may be licensed by multiple other jurisdictions. However, the general procedure to obtain each of the required licenses follows a similar pattern.

The licensing process imposes obligations on both the government agencies issuing the licenses and the operators who are the licensees. The license approval process typically includes an inter-agency coordination process in which the licensing authority consults with other government agencies who might be affected by, or have oversight of, the proposed operation. This reduces the administrative burden on the operator by reducing the number of consultations they must undertake. Licensing authorities often also have an obligation to conduct technical and financial due diligence on applications received. This helps reduce the number of frivolous applications received, and helps prevent resources (such as spectrum) from being allocated to operators who are unable to use it. Operators should be prepared to respond to due-diligence requests during the license approval process.

When applying for a license, operators should be aware of potential administrative fees and the time required to process the application. Fees are intended to allow the issuing authority to recover costs associated with processing the applications. Application processing times vary, but can be significant depending upon the efficiency of the authority and the amount of interagency coordination required. For applications requiring full coordination and processing with the ITU, the processing time required can be measured in years. System deployment plans must account for these processing times.

Licensing applications, processes, and requirements may differ by operating domain or type of system. Systems operating in geostationary Earth orbit (GEO) may be subject to a different process than those operating in other orbits, which may be grouped together as non-geostationary orbit (NGSO). In the telecommunications segment, Fixed Satellite Services (FSS), Mobile Satellite Services (MSS), and Broadcasting Satellite Services (BSS) can have differing licensing processes. Some national regulators may offer less onerous licensing requirements for amateur satellite operators, and some authorities responsible for launch operations offer distinctions between experimental and operational systems. It is the responsibility of the operator to ascertain which categories are applicable to their system, although national licensing authorities may offer consultation on this subject. During the application process, applicants should also be aware that some national regulators make applications public (either in total or in part) and may also allow public commentary on applications. This may present implications for business strategy of commercial license applicants.

> **Licensing applications, processes, and requirements may differ by operating domain or type of system.**

Once a license is issued, the operator is responsible for various continuing reporting requirements. Licenses typically have a validity period, after which a renewal application may be required. Satellite operators are commonly required to report any major changes in system operations or performance, significant operational anomalies, including technical faults, to the licensing authority, and may also be required to submit annual performance reports. These reporting requirements satisfy the licensing authority's national obligation under Article VI of the Outer Space Treaty to provide continuing oversight of licensees under its jurisdiction.

Launch Vehicle Selection

When selecting a launch service provider a range of technical and operational considerations must be evaluated to assess providers' and vehicles' suitability to perform the spacecraft operator's specific requirements. Some spacecraft operators may wish to engage a technical consultant to advise on launch vehicle selection. Generally speaking, most commercial spacecraft operators will seek proposals from multiple launch service providers.

If a satellite operator requires launch insurance (and most operators do require insurance in order to meet financial obligations and licensing requirements), an insurance broker will likely work with the satellite operator and the proposed launch providers to determine the appropriate insurance rates. If possible, the satellite operator should expect to have several of its own technical experts on-site at the launch facility during the launch campaign to participate in the integration of the satellite with the launch vehicle and other final preparations for the launch. Insurance is discussed later in this chapter. It is important to select a launch vehicle with adequate performance capability and appropriate performance margin to accommodate modest satellite mass growth if necessary. Launch service providers will not allow their limits to be exceeded because this will result in catastrophic failure or deployment into an incorrect orbit.

Launch providers normally have a queue of payloads waiting to be launched. Conducting a space launch is a complex endeavor requiring coordination of many complicated tasks that are affected by a variety of factors that are difficult to control. The launch vehicle and satellites are often composed of components manufactured by dozens to hundreds of suppliers. Those components must be tested to ensure proper function before and during integration between the satellite and the launch vehicle. Any anomalies discovered during testing often require disassembly and further testing. Furthermore, failure of a satellite or launch vehicle in orbit that shares hardware with a new satellite in manufacture may require a delay in production until the cause of the other mission's failure is determined. Even if a spacecraft and launch vehicle show up at the launch site on schedule, it may be necessary to wait for the launches of other payloads that have priority but have experienced schedule slips. Once on the launch pad, factors such as weather, launch range, and possible conjunctions with objects already on orbit, can further delay a launch. All of these factors lead to the reality that many launches do not occur when originally scheduled.

Multiple Payload Launch and Deployment Options

Multiple choices have emerged for how to deploy payloads into their desired orbit. The traditional option of launching a single payload on a dedicated rocket directly into its required orbit remains popular with larger and more expensive satellites, but often comes with a significant price tag. The growth of large constellations of small satellites has also led to an increase in clustering, where a single rocket is loaded up with multiple satellites, sometimes dozens, from the same operator that are then deployed into the same orbital plane at the same time. Subsequent launches deploy more clusters into additional planes, thereby filling out the desired constellation design over time.

The growing demand from newer actors who want to launch smaller satellites has opened up new markets for rideshares to space, where many satellites are placed on the same rocket, sometimes from multiple customers. This often leads to less choice of which orbit any individual payload can go into, although more advanced restartable upper stages may give more flexibility to deploy into multiple orbits.

In some cases, a rideshare opportunity exists when the primary satellite weighs less than the rated capacity of the launch vehicles, creating the opportunity for other smaller payloads to ride along as secondary payloads at a reduced price. Entities wishing to pursue launch in a ridesharing arrangement might contract directly with a launch operator or work through a launch broker service, which matches payloads to launch opportunities. Some launch brokers may themselves purchase a dedicated launch opportunity and aggregate multiple payloads together. In still other cases, a launch operator may dedicate an entire launch vehicle to rideshare opportunities on a regular basis, such as quarterly, creating opportunities for rideshare payloads to manifest for the vehicle closest to when they are ready.

Figure 3.4 | Vega's Small Spacecraft Mission Service dispenser with multiple satellites is lowered onto the rocket's payload adapter at Guiana Space Centre in Kourou, French Guiana in June 2020. *Credit: ESA*

Ridesharing arrangements are typically lower in cost than purchasing a dedicated launch, which may be cost-prohibitive for many new actors. However, the approach has its drawbacks. Secondary payloads typically have a reduced ability to influence the schedule of the launch, which is usually negotiated between the launch operator and the primary payload operator. Secondary payloads may also find themselves with limited orbital insertion options and facing a suboptimal vibration and acoustic environment during launch, as these parameters are defined according to the mission requirements of the primary payload. Furthermore, a rideshare increases the complexity of the launch and deployment and therefore increases the risk of failure. Potential rideshare hazards that a satellite may experience as a consequence of being one of multiple payloads on a launcher include explosive hazards, electromagnetic compatibility, electrical shock, battery rupture, electrolyte leakage, sharp edges, protrusions, and premature mechanism deployment.

Some of the orbital insertion drawbacks are being mitigated by new in-space transportation and deployment services, which usually consist of a separate maneuverable vehicle called a deployer or orbital transfer vehicle. The deployer acts like an additional upper stage and can carry multiple payloads of its own from the orbit it was placed into by the main rocket to one or more new orbits that better suit its customer's needs.

Yet another option for launching a satellite is in-space deployment of satellites from other orbiting space objects. Crewed orbital platforms such as the International Space Station have also become "launch vehicles" of a sort. Satellites are carried up to the station via periodic cargo shipments and then deployed by astronauts through airlocks and using robotic deployment systems.

Figure 3.5 | Robotic deployment of satellites, in this figure from the Small Satellite Orbital Deployer (SSOD) held by the ISS's Kibo robotic arm, presents other opportunities to launch satellites into orbit. *Credit: NASA*

For some satellite missions, it may be more efficient to deploy a payload into space on another operator's satellite, a technique known as a hosted payload, thereby negating the need to build and launch a dedicated satellite. In a hosted payload configuration, the payload owner pays the host spacecraft operator to carry an instrument that uses the host satellite's utilities, such as power, data transfer, etc. Finally, as new, large constellations of communications satellites have been announced, a concept called a hosted bus has emerged. In this configuration, a satellite operator can purchase a spacecraft based on the same bus as the other satellites in the constellation. The hosted bus operator benefits because the nonrecurring engineering costs of the satellite bus have been paid for by the constellation operator, making the hosted bus satellite much less expensive to build. Another major benefit is that the hosted bus operator can use the constellation's communications network and ground infrastructure, and may be able to ride-share a launch for a relatively low price.

Launch Services Agreement

Securing a launch to outer space with a launch provider will require entering into a legally binding contract called a launch services agreement. The launch services agreement will methodically define all the particulars of launching, and give definitions for many elements of the launch. The agreement delineates all the particular roles and responsibilities of the actors, but in general, these are that the customer will be handing over a satellite that is fit for launch, and the launch provider will be performing certain services, such as successfully integrating the satellite into the rocket and safely and successfully launching it into the correct orbit.

Each launch services agreement will include unique elements for each particular launch, but—as with most contracts—it will always have certain elements that make it sufficient as a legally binding contract. While the contracts that companies use may seem lengthy, deal in minutiae, and address scenarios that might not happen (such as launch failures and other mishaps), legal contracts are nuanced documents that refine all of the various shared understandings and expectations of the parties into a finite number of words that address all details, define all roles, assign risks, and do so in a fashion that would stand up in court as being a valid contract (i.e., a written reflection of the parties' shared understanding of what they undertake to do).

So that both the launch provider and the customer have the exact same understanding of particular words, a launch contract will define its most important terms. The definitions section of a contract might define the following: "satellite," "launch services," "launch opportunity," "launch vehicle," "launch window," "launch" or "launching," "post-launches services," "shared launch," "third party," "auxiliary payload," "launch abort," "launch failure," "partial failure," and other important terms. Because they are defined, each party is held to understand these terms, and to agree to them upon entering into the contract.

For example, "launch failure" might be defined differently than "partial launch failure," and should the satellite not be placed into the correct orbit, the resulting situation might be categorized as a launch failure—or perhaps only a partial launch failure. This categorization might have a direct impact on the triggering of insurance and even liability provisions. The definitions in the launch contract matter, and should be deeply scrutinized by the parties.

Another component of the launch services agreement are the sections listing the undertakings to be executed by both sides. Sometimes called commitments, or technical commitments, these enumerate precisely what each side must do so that the other side can fulfill its obligations under the contract. Because launching advanced hardware to outer space is such a technological achievement, the parties are essentially becoming partners with each other for a certain amount of time. Launch service providers are also typically responsible for verifying payloads being launched have secured all necessary licenses or authorizations prior to launch, including payload safety reviews, necessary spectrum or remote sensing licenses, or any national registration reporting requirements. Payload operators' failure to secure these required licenses in a timely manner might result in removal from a launch manifest.

Lastly, parties to a launch contract must face the possibility of disaster, and consider, negotiate, and agree upon what risks are borne by whom, what rights are accorded in the case of certain events, and what roles each party must play. A section of the contract will contain some allocation of potential liabilities and risks.

Standard contracts outside of the space industry have a clause sometimes called a "force majeure" clause, which means that an intervening, supervening, or otherwise unpredictable "act of God" will excuse the parties from undertaking their commitments under the contract.

Insurance

Insurance may be required by the national regulatory authority licensing and supervising the space activities. It might also be required by the launch services provider in the launch services agreement. A launch buyer may procure insurance to minimize exposure resulting from a launch failure. Generally, launch vehicles with a less reliable track record have more expensive insurance while more reliable systems have less expensive insurance. Therefore, insurance can balance out the price differential between low-price, high-risk launch options and high-price, more reliable launch providers. Operators commonly purchase insurance for launch, which provides coverage from launch-vehicle ignition to on-orbit delivery. Operators may also purchase a policy which covers launch plus a certain period of time for initial operations (up to a year). A separate policy, if required, is purchased to cover satellite failure during its operational phase in orbit. A launch buyer should also be aware of the liability environments in the nations hosting the launch providers. If a launch failure causes damage to the uninvolved public, a launch service buyer may be exposed to liability. Some countries have put in place indemnification regimes that establish a maximum third-party liability level so that damages in excess of that amount are paid for by the national government. Not related to launch, but some operators in GEO may purchase specific policies to cover the spacecraft during maneuvers to relocate between different slots.

Launch Mission Assurance

Launch operations deploying satellites rely on a partnership between the launch operator and the launch buyer to implement a process and culture focused on mission success. This type of relationship and process, called mission assurance, is a standard that is perhaps not feasible for smaller commercial budgets, but can be employed by large-scale buyers, such as national governments. Mission assurance as a process is an iterative and continuous technical and management activity employed over the entire lifecycle of a launch system. To achieve success, the mission assurance process must include a disciplined application of systems engineering, risk management, quality assurance, and program management principles.

Key features of mission assurance include a launch procurement strategy that includes adequate contingency funding, which then ensures that the launch provider maintains the workforce, facilities, and data-sharing required to perform integration and launch, handle contingencies, and reach agreement when issues arise. Another key feature of mission assurance is clear accountability, which requires that a single entity is responsible for understanding, tracking, and ensuring that flight worthiness is maintained.

Next, continuity and independent verification require that funding is available to maintain the depth of independent technical capabilities to analyze potential issues and render assessment of spaceflight worthiness. Finally, it is necessary to conduct extensive reviews; both those leading to the spaceflight worthiness certification and the go/no-go decision for launch, as well as post-flight data reviews.

LAUNCH

The launch phase is considered the riskiest time period during any space project. Significant care must be taken to both increase the odds of a successful space launch and minimize the risk that space launch activities pose to people, ground installations, and air and maritime vehicles.

There have been satellite launches from more than three dozen sites around the world. Today, most launches occur from around twenty launch facilities, although there are plans by many countries and companies to increase that number in the near future. Creating and safely operating a launch facility requires thorough consideration of launch safety, environmental, and ground safety issues. Spaceports are generally located in sparsely populated regions to minimize the risk that a launch failure could harm people or property in the area. Spaceports are also often located near oceans or deserts so that a rocket's ascent trajectory overflies large, relatively uninhabited regions in order to minimize public exposure to expelled rocket stages or other falling debris. Once a site for a launch facility is identified, a national government often requires completion of both a scoping and a comprehensive environmental impact assessment that assesses the environmental, ecological, socioeconomic, cultural, and other impacts of the operation of a launch facility at the proposed location. Finally, the design and operations of a spaceport need to follow best practices that have evolved at established spaceports.

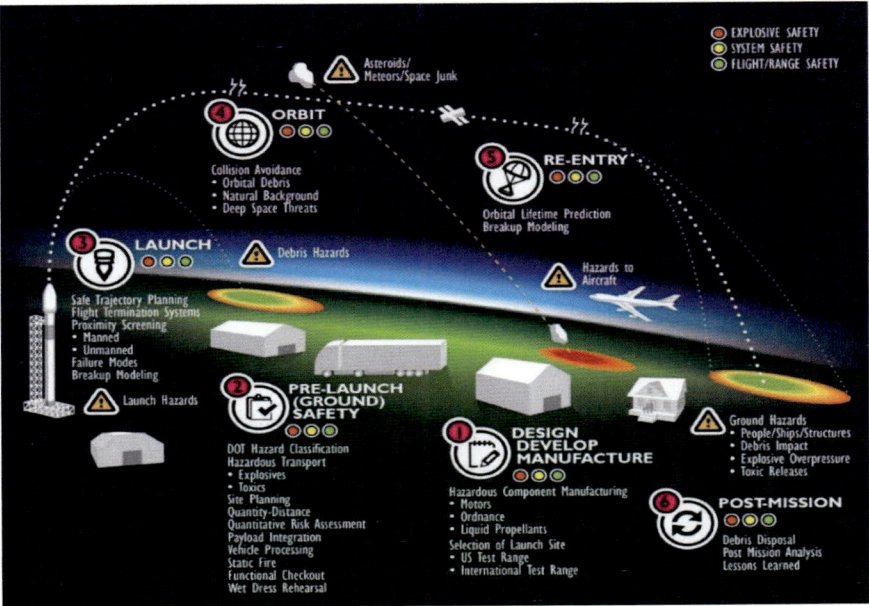

Figure 3.6 | This figure illustrates the typical stages a launch vehicle goes through and the types of safety considerations at each stage. *Credit: FAA*

There are no globally agreed-upon rules for how to develop and operate a space launch facility. Spaceports have traditionally been developed as national assets and managed by government agencies, although an increasing number of private sector entities are becoming active in spaceport development and operations. Within the United States, many states have conducted studies to determine a path forward toward their own local commercial spaceport development. Some U.S. states have taken steps to incentivize and enable development of commercially operated spaceports. The U.S. federal government, through the Federal Aviation Administration (FAA), has created the most proactive commercial spaceport regulatory regime thus far, and other countries often reference U.S. regulations. However, many of these proposals for state or regional spaceports have faced challenges from local public opposition and in building a business case, and few are actively generating revenue.

Terrestrial Environmental Safety Considerations

The terrestrial environmental impact of constructing and operating a proposed launch site may be significant, and the relevant national authority will likely require an environmental impact analysis. Developers of launch facilities need to take into account the effect of launch activities on various environmental domains including the atmosphere, noise sources and effects, and surface environments.

One environmental concern is the impact space launches have on the atmosphere. Industrial emissions are often regulated by national air quality standards to ensure that pollutant levels do not exceed certain regulatory thresholds set to protect public health. Due to their ultrahazardous effect on ambient air quality if they are accidentally released, the storage and use of some high-energy and volatile rocket fuels may be of unique concern. In addition, some launch vehicles emit hazardous gasses during normal operation. Other types of launch vehicles, especially those with solid rocket motors, emit various types of particles when traveling through the upper layers of the atmosphere, which may come under increased scrutiny by environmental regulators in the future.

A second major source of environmental concern is noise. The amount of noise created by a proposed launch facility needs to be understood and evaluated in the context of the natural noise environment. Rocket launches tend to generate significant amounts of noise that can disrupt wildlife habitats. Sonic booms generated by launch and re-entry activities along a trajectory may cause further damage to wildlife, property, and human physiology.

Finally, launch facilities are often placed in areas that are remote from human populations, but may also be pristine wildlife habitats. Land, marine, wetland, and other surface environments surrounding a launch site may each have unique features requiring protection. Site-specific studies and impact mitigation plans should be in place prior to construction. Developing a launch facility near areas containing threatened and endangered species habitats should be especially avoided.

Ground Safety Considerations

Once appropriate environmental concerns are addressed, a national regulatory entity will likely require a policy review to ensure that a proposed new space launch facility would not jeopardize national security, foreign-policy interests, or international obligations of the hosting nation.

Next, a casualty risk assessment will be conducted, where casualty refers to the number of people who are predicted to be killed or seriously injured by a launch accident. Launch sites should be placed in areas where launch activities will not jeopardize public health and safety or the safety of property. Therefore, the flight corridor for a launch vehicle—the land under its launch trajectory—must be adequately unpopulated so that there is a minimum chance of damage should the rocket vehicle or spent stages impact the area. Models exist to calculate the risk to the public, and some nations, such as the United States, set maximum acceptable quantitative casualty risk levels.

Because of the explosive nature of many solid and liquid propellants, another key part of the initial design of a space launch facility is the creation of an explosives site plan that shows the location of all explosive hazard facilities, the distances between them, and the distances to public areas. Safe handling and management of explosive launch-vehicle propellants is critical. Standards exist to guide construction of launch site infrastructure in order to avoid causes of accidental

Case Study: SpaceX Starbase

The tensions among spaceport activity, wildlife habitat, and economic interests were demonstrated in the ongoing debates about SpaceX's Boca Chica Launch Site in Texas. Known as Starbase, this commercial spaceport is located near the Gulf of Mexico and adjacent to a sensitive wetlands area and is home to several endangered or protected species. Plans to expand its operations to include the launch of Starship lead to concerns about the environmental impact.

Figure 3.7 | Photo taken from the ISS shows SpaceX Starbase and surrounding areas in southern Texas. *Credit: NASA*

The U.S. Federal Aviation Administration (FAA) performed a lengthy Programmatic Environmental Assessment regarding the SpaceX proposal to carry out its Starship/Super Heavy Launch Vehicle Program. The FAA found that Starbase's operations would not have a significant detrimental impact on the environment and the public, but also required SpaceX to implement dozens of mitigation measures to ensure that was the case. These measures ranged from providing advanced notice of road closures to changing lighting in the facility to providing public notifications about noise and other factors.

explosions, such as lightning, static electricity, electric supply system problems, and electromagnetic radiation.

To ensure safe space launch facility operations, it is important for an operator to address controlling public access, scheduling operations at the site, notifications, recordkeeping, and launch site accident response and investigation. Access to the site should be controlled using security guards, fences, and other barriers. People entering the site should be taught the safety and emergency response procedures. Alarms and other warning signals are necessary for informing people at the site of an emergency situation. If a launch site has multiple users on the site at the same time, the site operator should have procedures for scheduling operations so that the activities of one do not create hazards for the other.

Hazard areas are another particular concern. Coordination with the national maritime and air traffic control entities is necessary to limit how closely aircraft and watercraft can approach launch and re-entry operational hazard areas. Notices to Mariners are issued for spaceports near waterways when launch activities are being conducted. The notices require vessels to clear hazard areas during specific windows of time. Alternatively, Notices to Air Missions (NOTAMs) are issued for areas surrounding a launch facility and beneath a launch corridor when expected casualty calculations exceed specified thresholds. When a launch facility conducts a flight operation, the appropriate equipment to track a launch vehicle's progress across the launch range must be aboard the launch vehicle and on the ground.

Range Safety During Launch Operations

The launch of a satellite requires significant planning, coordination, and risk management. Range safety operations have evolved over time at launch facilities around the globe. Standards that are in development by the International Organization for Standardization (ISO) identify safe practices that apply to launch site operations, flight safety systems, and other areas. Globally, most spaceports are operated by national governments and have varying approaches to the specific range safety practices. However, core principles are common. Range safety practices discussed in this section most often reference the commercial regulations developed and implemented by the U.S. FAA.

First, a flight safety analysis is conducted by a launch operator for each launch in order to control the risk to the public from hazards created during both a normal and a malfunctioning launch-vehicle flight. A risk assessment analysis should account for the variability associated with each source of hazard during flight, the normal flight and each failure response mode of the launch vehicle, and each external and launch-vehicle flight environment. Additionally, a risk assessment should consider populations potentially exposed to the flight, and the performance of any flight safety system (including time delays associated with the systems). The outputs of a risk assessment are used to create a plan to sufficiently isolate the hazard to keep risk to the public within acceptable quantitative limits. For example, the FAA provides the Flight Safety Analysis Handbook for reference in this process.

Types of Analyses	
» Trajectory	» Hold-and-resume gate
» Toxic release hazard	» Straight-up time
» Overflight gate	» Ground debris risk
» Flight safety limits	» Flight hazard area
» Probability of failure	» Orbital debris
» Time-delay	» Collision avoidance
» Malfunction turn	» Data loss flight time and planned safe flight state
» Far-field overpressure blast effects	

Table 3.1 | An example of the various analyses required as part of a full flight safety analysis.

Public Risk Criteria

National regulatory entities, such as the U.S. FAA, set specific quantitative criteria for the risk exposure of the public that launch operations must meet. These standards consist of specific probabilities of risk to the public from inert and explosive debris, toxic release, and far-field blast overpressure. These quantitative limits do not apply to aircraft or watercraft, and as a result, a launch operator must establish hazard areas with rules requiring the removal of waterborne vessels and aircraft from the hazard zone during the launch activity.

Flight Termination System

In order to meet public risk criteria, it is necessary to incorporate self-destruct systems on launch vehicles. Activation of a destruct system breaks the launch vehicle into smaller debris, burns off fuel, and keeps overpressure effects isolated from the public. Termination criteria are developed during various flight safety analyses and implemented as part of the written flight safety plan. Flight termination systems are a critical element of range safety. There are some exceptions to this rule, especially in the older rocket systems that use toxic fuels, in which case it is preferable for the rocket to destruct farther from the launch site on a trajectory that is routed into unpopulated areas.

Flight Safety Plan

Based on the conclusions reached during the flight safety analysis, a written flight safety plan defines how launch processing and flight of a launch vehicle will be conducted without adversely affecting public safety and how to respond to a launch mishap. A flight safety plan should identify the flight safety personnel who will approve and implement each part of the plan.

Elements of a flight safety plan include flight safety rules, a flight safety system, data on trajectory, and debris dispersion data. The plan must also identify flight hazard areas that must be cleared and controlled during launch, and support systems and services including any aircraft or ship that a launch operator will use. Last, the plan must have a description of the flight safety-related tests, reviews, rehearsals, and other safety operations.

A ground safety plan describes the implementation of hazard controls that have been identified by a launch operator's ground safety analysis and that address all public-safety-related issues. The plan should at least include a description of the launch vehicle and any payload (or class of payload), and identify each hazard, including explosives, propellants, toxics and other hazardous materials, radiation sources, and pressurized systems. The plan must also include figures that show the location of each hazard on the launch vehicle and indicate where at the launch site a launch operator performs hazardous operations during launch processing.

A VARIETY OF OTHER PLANS ARE NECESSARY
AS PART OF A FLIGHT SAFETY PLAN, INCLUDING:

» *Launch support equipment and instrumentation plan*

» *Local agreements and public coordination plans*

» *Frequency management plan*

» *Hazard-area surveillance and clearance plan*

» *Flight termination system electronic piece-parts program plan*

» *Communications plan*

» *Accident investigation plan*

» *Countdown plan*

Safety-Critical Preflight Operations

A launch operator must perform safety-critical preflight operations that protect the public from the adverse effects of hazards associated with launch processing and the flight of a launch vehicle. For example, a launch countdown plan should be distributed to all personnel responsible for the countdown and flight of a launch vehicle. Any nearby region of land, sea, or air necessary to the launch should be assessed and monitored to ensure the number and locations of members of the public meet established safety standards. The operator should monitor the weather to identify meteorological conditions that could threaten the safe performance of a launch, such as the presence of lightning. To ensure accuracy, data verification of launch-vehicle tracking should be employed.

If the launch vehicle exits flight boundaries, the readiness of flight safety systems must be ensured if intentional destruction of the launch vehicle is required. At least two tracking sources should be available prior to liftoff, and no less than one verified tracking source at all times from liftoff to orbit insertion for an orbital launch, or to the end of powered flight for a suborbital launch.

Launch Vehicle Stage Disposal

Launch vehicle stages pose hazards for spaceflight and environmental safety if not properly disposed of. Upper stage launch vehicle stages left in orbit after launch operations are complete become space debris, while upper stages that conduct uncontrolled re-entry pose a risk to safety of third parties on Earth. The best way to mitigate both of these hazards is to conduct a controlled, planned re-entry disposal of upper stages, targeting unpopulated areas of the Earth (generally open ocean). Other methods of disposal include maneuvering the stage into an approved "graveyard" orbit, sending the stage on an Earth-escape trajectory, or retrieving it via active debris removal service. If the stage must be left in orbit, it should be passivated and placed in a trajectory that naturally re-enters Earth's orbit in compliance with the 25-year "rule" and in a way that ensures maximum burning up in the atmosphere. Launch vehicle lower stages do achieve orbit, and thus do not pose long-term space debris risk. However improper disposal of lower stages can pose environmental and safety risks on Earth. Responsible range safety provisions are typically designed to ensure lower stages do not pose casualty risk to down-range populations. Lower stages may contain hazardous fuel or other materials that if not mitigated may cause harmful effects. Use of nonhazardous or recovery and reuse of lower stages can be strategies to reduce this potential harm.

IN–SPACE ACTIVITIES

Remotely operating spacecraft to ensure mission assurance and safety of flight requires managing a variety of risks—not the least of which is avoiding running into other active satellites and the hundreds of thousands of pieces of space debris also orbiting the Earth. The following sections provide a more detailed discussion of the major issues that satellite operators need to deal with in order to ensure the well-being of their satellites and prevent collisions or incidents that could undermine the long-term sustainability of space activities.

Satellite Orbit Determination and Tracking

The first step is for satellite operators to be able to know where their own satellite is in orbit, and know the locations of other objects that may pose a collision risk. While a growing number of satellites have onboard global satellite navigation system (GNSS) receivers to determine their position, there is still a need for alternate ways to validate the GNSS solution and also track the tens of thousands of nonfunctional objects. The majority of space objects are debris objects and must be observed using systems which do not rely on the cooperation of the object being tracked in order to determine their orbit.

Satellite operators need to determine how they will obtain orbital trajectory information on their satellites and other space objects. Satellite orbit determination (OD) is the process by which operators or third parties can obtain knowledge of the satellite's trajectory, usually relative to the center of mass of the Earth. The basic theory involves determining a satellite's position and velocity—its state—at a specific time in the past, and then using a set of differential equations that model changes in its position and velocity over time to predict where it will be in the future. In aerospace terms, this is "generating an ephemeris," which is a set of points in space that define the future trajectory of a satellite. A significant challenge in performing accurate OD is developing precise and accurate equations of motion that include the various natural forces or perturbations that act on the satellite, such as irregularities in Earth's gravity, atmospheric drag, and the gravitational pull of the sun and the moon.

Remotely operating spacecraft to ensure mission assurance and safety of flight requires managing a variety of risks—not the least of which is avoiding running into other active satellites and the hundreds of thousands of pieces of space debris also orbiting the Earth.

Satellite OD begins with data on the position and velocity of a satellite, known as observations. A single observation measures a satellite's position, and perhaps velocity as well, at a specific moment in time, and relative to the location of a specific sensor. Multiple observations taken over a single period of time are called a track. The observations from one sensor can be used by themselves or combined with data from other sensors which observe the space object at other points in its orbit.

Different measurement types have different characteristics, and these lead to different levels of confidence in satellite state elements estimated from these measurements. Traditionally, the main source of data has been collected by ground-based radars and ground- and space-based telescopes. Telescopes may also use satellite laser ranging (SLR) techniques to directly illuminate satellites using a laser source to obtain range information. Radar observations can provide velocity information and typically have excellent angular tracking, but can suffer from poor range rate estimations. SLR can derive excellent range and range rate estimations while having poor estimations of angular rates.

No matter the type of sensor, it is important to understand the accuracy and precision of the tracking data it provides. Often, sensors are periodically tasked with tracking calibration spheres or other space objects whose orbit is well known in order to determine their accuracy. If a sensor's measurements are consistently off-true, a deliberate bias can be introduced to correct some or all of the error. The historical performance of sensors can be tracked in order to determine their accuracy and precision over time, which in turn can be used as a weighting factor for valuing their data relative to other sensors.

Accurately tracking a space object requires collecting observations from many parts of its orbit. That means a global network of sensors is required, which can be terrestrial or space-based. To operate and maintain such a network has historically been expensive, and as a result, tracking satellites and space debris has been primarily a governmental function. To date, the U.S. military has been the primary source of this type of information to the public, although there are increasingly other sources of tracking information (both governmental and non-governmental) that are becoming available to satellite operators.

Orbit Propagation

Knowing where an object is now, however, is only part of the problem, since there is also a need to know where an object will be in the future to assess the risk of collision. That means understanding the various forces acting on an orbital object—Earth gravity, solar and lunar gravitational effects, solar radiation pressure, and atmospheric drag, the last of which presents a major challenge for LEO objects. Much scientific research has gone into developing mathematical models to estimate how these and other natural forces—known as perturbations—affect satellite trajectories over time. But one force can be extremely difficult to model: the thrust used to maneuver a spacecraft. Most active spacecraft have to maneuver periodically to maintain the orbit needed to perform their mission. A maneuver that takes place during the timeframe of a future prediction—such as the probability of whether the satellite will collide with another object—will invalidate the prediction from the moment that the maneuver is conducted. Thus, accurate modeling and predictions need to take into account both models of natural perturbations and any planned maneuvers.

The good news is that the satellite operator must know this information well in order to perform their mission. Sharing the information with other operators can provide more timely updates and avoid confusion as a result of not knowing an operator's intentions. The challenge is that each operator typically uses their own coordinate systems (and sometimes different time systems), which means they all have to be normalized—or put in a common reference system—to be useful. This process requires a full understanding of units, coordinate and time definitions, and a way to validate that information, since many satellite systems were not designed to interoperate with those of other operators, only to be internally consistent.

The results also need to be shared in a standard way to ensure that each operator knows how to understand and apply that normalized data. And that sharing needs to be done on a regular basis to ensure a common understanding of how to apply the data and to avoid the possibility of misinterpretation in the midst of responding to a serious event.

Conjunction Assessment Procedures and Standards

For a satellite operator, one of the key tools for reducing in-space risk is to perform conjunction assessment (CA); that is, to determine which objects might have a chance of coming close to, and possibly colliding, with your spacecraft. Conceptually, the CA task is straightforward. The operator simply needs to know where all the objects are that might present a collision risk, and be able to predict where they will be for a period far enough into the future to enable an effective course of action should a close approach be deemed unsafe. With that information, the process of screening each of the operator's satellites can be performed quickly using well-known analytical techniques. The challenge comes from understanding current limitations to performing effective CA and identifying areas for improvement. The NASA Spacecraft Conjunction Assessment and Collision Avoidance Best Practices Handbook provides further in-depth examination of these issues.

Typically, CA is performed for a pair of trajectories, each representing the location of a space object over time, where the relative separation distance between two objects is computed over a given prediction time span. The trajectories may be generated using high-accuracy catalog data from a data provider, or using positional data generated by the spacecraft itself. A conjunction event is where the relative separation reaches a local minimum, commonly referred to as the point of closest approach.

Operational collision risk management starts with the generation of close approach predictions and ends with an action/no-action decision from mission stakeholders. The step-by-step process consists of:

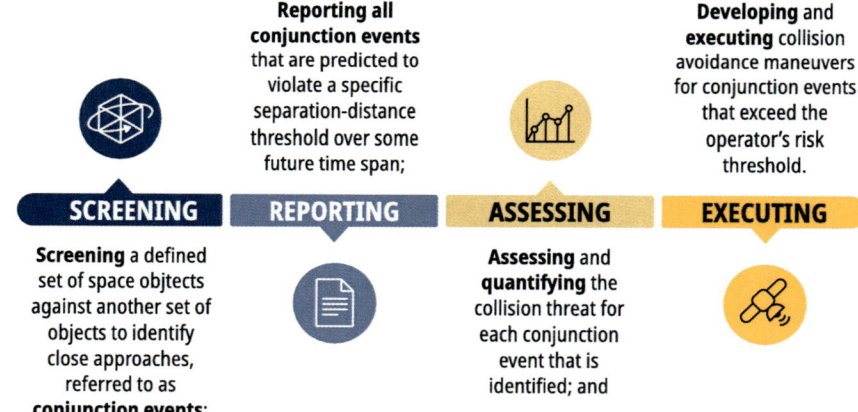

SCREENING — Screening a defined set of space objects against another set of objects to identify close approaches, referred to as **conjunction events**;

REPORTING — **Reporting all conjunction events** that are predicted to violate a specific separation-distance threshold over some future time span;

ASSESSING — **Assessing and quantifying** the collision threat for each conjunction event that is identified; and

EXECUTING — **Developing and executing** collision avoidance maneuvers for conjunction events that exceed the operator's risk threshold.

Figure 3.8 | Operational collision risk management process.

Potential collisions can be identified by individual spacecraft operators, operational support organizations such as the Space Data Association (SDA), government organizations such as the European Union's Space Surveillance and Tracking (SST) Program, and a growing number of commercial providers. To be most useful to satellite operators, the entity conducting the conjunction analysis should have accurate trajectory data on both active satellites, including planned maneuvers within the prediction time, and other space objects.

Operational Conjunction Assessment

The conjunction assessment process occurs throughout the lifetime of a satellite, from pre-launch to end-of-life operations. Phases of conjunction assessment include launch, early orbit, on-orbit, collision avoidance, and de-orbit or disposal. Launch conjunction assessment is the process of predicting and reporting the close approaches between launch vehicles and orbiting objects. This is done by screening planned launch trajectories against all objects in the space catalog. The launch provider typically generates the trajectories, which may include multiple iterations corresponding to different launch times within the launch's window of opportunity.

The process of launch screening compares the trajectory of the launch vehicle (delivered as ephemeris data) to a catalog of space objects. The preliminary screening process may begin weeks to days ahead of the launch date depending on the launch provider's or launch range's requirements. Subsequent screenings are then performed at predetermined intervals, such as at T-4, 3, and 2 days before launch, and finally on the day of the launch, to produce the most accurate and timely assessment.

Screening results are provided for predetermined screening volumes that depend on the satellite mission. For example, a robotic mission with active payloads may use a stand-off screening distance of 25 kilometers. This means that the launch operator will be notified of any predicted close approaches with miss distances less than that.

A number of entities provide launch conjunction assessment services. The U.S. military performs launch conjunction assessment for all launches that occur from the U.S. Air Force's eastern and western launch ranges, as well as for any other global launch provider who requests the service. Other data providers, such as Aerospace Corporation, also provide launch conjunction assessment, and many launch agencies across the world perform independent internal assessments using publicly available data.

There is ongoing debate about the usefulness of pre-launch conjunction assessment. In many cases, there is a significant amount of uncertainty in the predicted insertion orbit and the predicted trajectories of existing satellites. As a result, launch conjunction assessments may yield a high degree of false positives, and may unnecessarily cause launch delays or aborts. Some launch operators have concluded that it is only worthwhile to conduct launch conjunction assessments against the ISS, while others do so for a much larger number of satellites and debris objects. However, one significant benefit of conducting launch conjunction assessment screenings is that a satellite operator will discover which other objects are "in the neighborhood," and thus, with which other operators they will need to establish working relationships. In some cases, satellite operators have decided to modify the planned operational orbit for their satellite based on a launch conjunction assessment which showed that it was going into a high-traffic region. In the case of China's TanSat, the decision was made to not launch it into the "A-Train" constellation of Earth observation satellites due to the complicated requirements and procedures necessary for all participants in the A-Train.

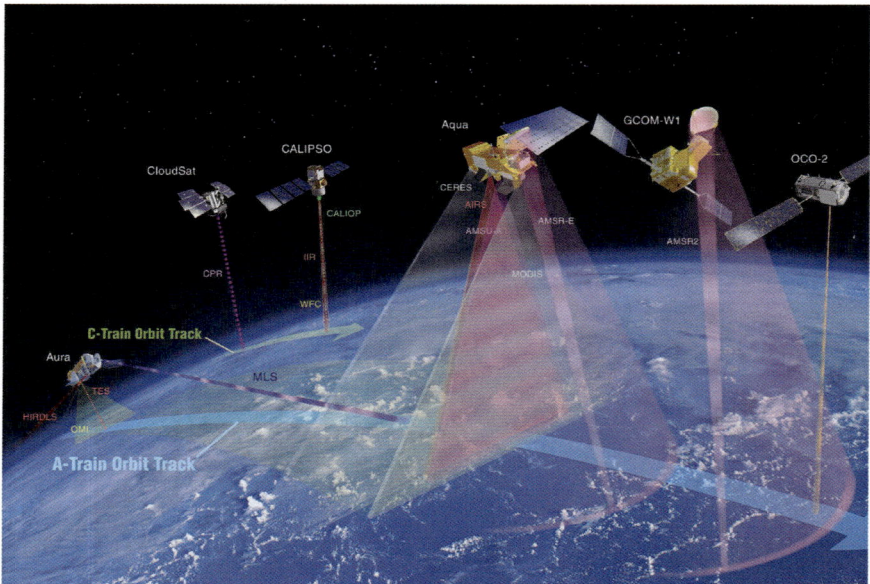

Figure 3.9 | Depiction of NASA's Afternoon Constellation ("A-Train" and "C-Train"). *Credit: NASA*

Early-orbit conjunction assessment spans the phase from the spacecraft's separation from the launch vehicle to its arrival at its final orbit. This phase can take days or months depending on the maneuver plan and methods, and presents unique challenges to the conjunction assessment process. First, the limited observational data in the first few days after launch can delay the ability to generate an accurate prediction of a newly launched object's future trajectory. Additionally, the spacecraft's constant maneuvering makes it difficult to maintain consistent tracking and updated orbit determinations. Consequently, accurate and timely early-orbit conjunction assessments often require the use of operator-provided data for ephemeris-based screenings.

Early-orbit conjunction assessment typically includes the operator providing the early-orbit maneuver plan to a data provider in addition to a schedule of planned maneuvers and required screening volumes. As the early-orbit phase progresses, the operator provides the given satellite's ephemeris to the data provider for pre- and post-maneuver screenings against the space catalog. This data exchange allows the operator to perform collision avoidance, if needed, and helps the data provider maintain accurate positional data for the maneuvering satellite. The Combined Space Operations Center (CSpOC) provides this service for free to all satellite operators who provide their ephemeris, while operators can request European Union Space Surveillance and Tracking (SST) Services for Collision Avoidance, which is also free and open to non-EU users. Some space agencies also provide the service for their own governmental payloads.

Case Study: Europe's Sentinel 1-A Satellite

A real-life situation where early-orbit conjunction assessment created challenges involved Europe's Sentinel 1-A satellite. Sentinel 1-A was launched on April 3, 2014, and within its first day on orbit, it was predicted to have a very close approach with a defunct American satellite which had not shown up during the pre-launch screening. Planning and conducting the maneuver proved to be very challenging, as Sentinel 1-A was still in the process of conducting a set of maneuvers to deploy its solar arrays and antennas. Ultimately, the maneuver went smoothly and a potentially disastrous situation was avoided.

Figure 3.10 | ESA illustration of the Sentinel 1-A C-band Synthetic Aperture Radar (SAR) satellite. *Credit: ESA*

Several commercial companies have started to offer SSA data and services. However, as is the case with launch conjunction assessments, early-orbit conjunctions can be difficult to predict in advance.

In-space conjunction assessment is primarily used to ensure spaceflight safety throughout the lifetime of a satellite. The process screens all active satellites against all other cataloged space objects. The results provide satellite operators with predictions of future close approach events. The close approach prediction information allows satellite operators to take actions to mitigate the risk of collision. The primary metric for doing so should be the probability of collision (P_C).

Close approach screening results are performed for prediction times that are dependent upon the satellite's orbital regime. The prediction time for satellites in GEO is typically longer than that of all other regimes, largely because GEO orbits are more predictable over long periods. The screening volume also varies across the different orbital regimes, and often includes a larger monitoring volume and a smaller high-interest, or reporting, volume. Table 3.2 provides an example of how different orbital regimes may be defined and assigned specific screening durations and volumes depending on their level of risk.

Orbit Regime	Orbit Regime Criteria	Propagation Time	Radial Miss (km)	In-Track Miss (km)	Cross-Track Miss (km)
GEO/HEO/ MEO	1300 min < Period < 1800 min Eccentricity < 0.25 & Inclination < 35	10 Days	10	10	10
LEO 4	1200 km < Perigee < 2000 km Eccentricity < 0.25	5 Days	0.4	2	2
LEO 3	750 km < Perigee < 1200 km Eccentricity < 0.25	5 Days	0.4	12	12
LEO 2	500 km < Perigee < 750 km Eccentricity < 0.25	5 Days	0.4	25	25
LEO 1	Perigee < 500 km Eccentricity < 0.25	5 Days	0.4	44	51

Table 3.2 | Examples of CA screening volumes. *Credit: Unites States Space Command*

The satellite operator, data provider, or service provider may perform conjunction screenings based on schedules dictated by specific missions using any variation of trajectories, as described before. Currently, the U.S. government provides free basic CA screenings for all global space operators and may provide additional services upon request. Private sector service providers such as the Space Data Association specialize in ephemeris-versus-ephemeris screenings, a complimentary service for satellite operators who elect to join the organization. A growing number of commercial companies also provide screening services.

Conjunction assessment reports may be issued and exchanged in a variety of ways, but the prevailing standard is the Conjunction Data Message (CDM) that has been defined by the Consultative Committee for Space Data Systems (CCSDS), an international standards body of space agencies.

Risk Assessment and Avoiding Collisions

Not all satellites possess maneuvering capability, but for potential collisions that involve at least one satellite with maneuvering capability, decisions on whether to conduct maneuvers to reduce the risk of a collision must be made. The decisions involve calculating the risk of collision and the potential costs of a maneuver (such as expending fuel or disrupting operations). Calculating the risk of collision requires not just knowledge of where the two objects will be, but also the amount of uncertainty associated with that knowledge. The location and uncertainty give the probability of collision, which must be future-combined with the consequences of a particular collision scenario.

Unfortunately, just calculating the probability of a collision is difficult. Ideally, the CDM would include the uncertainty data, usually in the form of covariance, but that is not always the case. The problem is further compounded when tracking maneuvering satellites, since failing to recognize that a maneuver has occurred can create a bad orbital prediction, overinflated uncertainty, or both. Similar results can be seen when trying to process observations for GEO satellites operating in clusters when observations are incorrectly associated with the individual satellites.

From a practical perspective, it is incumbent upon each operator to do their best to track their own satellites, regularly calibrate their results against other data sources and share that data with other operators in as timely a fashion as possible. The predicted trajectory should include natural perturbations and previously planned orbital maneuvers, and new orbital estimates should be provided as soon as possible after performing a maneuver or incorporating or canceling a planned maneuver. That data should be provided in the form of ephemerides far enough into the future to allow sharing and analysis of the data in support of decision-making—that is, early enough to plan and conduct an avoidance maneuver, if it is deemed necessary.

In the face of missing, incomplete, or potentially misleading uncertainty information, it is imperative that a variety of orbital data sources be compared to assess a more realistic uncertainty of the relevant orbits. This process must be applied for every case—not assumed to be the same from case to case.

Although it is impossible to prevent all collisions, these steps can mitigate the probability of a serious collision that can completely disable a satellite occurring and thereby creating the next large piece of debris or generating even more small debris that jeopardizes the entire near-Earth orbital environment. Collaboration and data sharing—between satellite operators and between operators and tracking services—are key to success.

Large Constellation Management

A key current trend in space operations activities is the planning, deployment, and operations of multiple very large satellite constellations (both commercial and governmental) consisting of thousands to tens of thousands of individual satellites in the same or similar orbits. These constellations challenge existing space operations practices and regulation, and require updated approaches. In the interim, operators have been working to develop large constellation management practices, both individually to specific constellations and in cooperation with other operations. For example, commercial satellite operators Iridium, OneWeb, and SpaceX worked together in a process facilitated by the American Institute of Aeronautics and Astronautics (AIAA) to develop and release (in September 2022) a reference guide to Satellite Orbital Safety Best Practices.

> From a practical perspective, it is incumbent upon each operator to do their best to track their own satellites, regularly calibrate their results against other data sources and share that data with other operators in as timely a fashion as possible.

Specific large constellation management practices that will need to be developed and enhanced include:

- Increased CA and collision avoidance on launch assessment (COLA) screening requirements, both internal to a single operator's constellation and between satellites in a constellation and other space objects, will require constellation operators to implement automated collision avoidance systems;

- Publication of satellite positional data (satellite ephemerides), health, and maneuver plans;

- Procedures for communicating and coordinating with other satellite operators in the same or similar orbits; and

- Proper practices for satellite deployment and check-out, as well as design-for-demise and reliable post-mission disposal.

Case Study: Voluntary Best Practices for Space Safety and Sustainability

In recent years, significant efforts have been made by a number of industry, government, and civil-society groups to develop and promulgate guidelines, principles, and best practices for space sustainability and/or space operations. Perhaps highlighted by the trend towards deployment of large constellations and partially motivated by questions over the efficiency and efficacy of existing regulatory processes, these efforts seek to establish voluntary principles and aspirational best practices for space safety that can inform more formal regulations in the future.

These efforts cover a number of types of space operations, including launch, in-space operations, and end-of-mission disposal. Some efforts address emerging areas of space activities, such as lunar operations and in-space servicing, assembly and maintenance (ISAM). Examples of some of most developed of these efforts include:

CONFERS

An industry association of 70+ companies that is developing best practices and standards for satellite servicing. CONFERS has developed and published a number of recommended operating practices and principles for commercial rendezvous, proximity operations (RPO), and on-orbit servicing (OOS), including the *CONFERS Recommended Design and Operational Practices.*

THE EARTH AND SPACE SUSTAINABILITY INITIATIVE

A UK-based initiative which is developing Environmental, Social and Governance (ESG) space sustainability standards to inform the finance and insurance communities and policymakers.

THE ESA NET ZERO DEBRIS CHARTER

A document written collectively by a group of space agencies and industry actors, facilitated by the European Space Agency, which lays out a set of "high-level guiding principles and specific, jointly defined targets" intended to drive the space community towards stopping creation of space debris as a result of space activities by 2030.

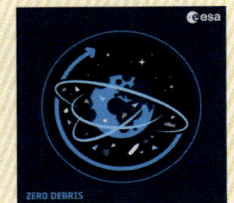

SPACE SAFETY COALITION (SSC)

An "ad hoc coalition of companies, organizations, and other government and industry stakeholders that actively promotes responsible space safety through the adoption of relevant international standards, guidelines and practices, and the development of more effective space safety guidelines and best practices." It publishes, coordinates, and updates a Best Practices for the Sustainability of Space Operations document.

THE GLOBAL SATELLITE OPERATORS ASSOCIATION (GSOA)

A trade association focused on the satellite industry, has published a Code of Conduct on Space Sustainability, which contains a set of recommended best practices to mitigate the risk of on-orbit collision, minimize the threat of non-trackable debris, protect humans in space, and limit the effects of satellite sunlight reflections on optical astronomy.

RESPONSIBLE OPERATIONS IN SPACE

These initiatives and others have resulted in a collective set of recommendations, principles, and guidelines which suggest there may be areas of emerging consensus on implementable best practices for safe operations. Some examples of key common principles expressed in many of these voluntary efforts include:

- Commitment to transparency and communication of intentions in operations
- Encouragement of intra-operator exchange of information relevant to safety-of-flight and collision avoidance
- Commitment to post-mission disposal within five years of end of mission
- Encouragement of design practices for responsible de-orbiting and passivation, including facilitation of servicing/removal
- Initial proposals for "rules of the road" for maneuver coordination and collision avoidance

However, as the various initiatives have developed somewhat independently of each other, more work needs to be done to establish areas of consensus, clarify how these voluntary practices should inform future regulatory regimes, and identify remaining gaps.

Many of the practices remain aspirational, and require validation and updating through operational experience. Implementation of these principles remains largely a voluntary commitment of individual operators, although in some cases individual documents have been formally developed into voluntary consensus standards, such as International Organization for Standardization standard *ISO 24330: Space Systems - Rendezvous And Proximity Operations (RPO) And On Orbit Servicing (OOS) - Programmatic Principles And Practices*, which was originally developed by CONFERS. Voluntary guidelines and practices might also be considered as part of voluntary space sustainability rating schemes—such as the Space Sustainability Rating—that could help to incent adoption and implementation of these voluntary standards and best practices. In some cases, elements of these voluntary best practices might ultimately be codified or reflected in binding regulation.

Space Weather

In addition to possible collisions with other space objects, the space environment itself can also pose a hazard to satellites. "Space weather" is the term for the set of physical and electromagnetic processes and effects that occur on the sun, and ultimately interact with the Earth's magnetosphere, atmosphere, and surface. These phenomena, which include solar flares, solar wind, geomagnetic storms, and coronal mass ejections, can have adverse effects on activities in orbit and on the Earth's surface.

The sun is constantly emitting electrically charged particles, which flow outward throughout the solar system in a phenomenon known as solar wind. The sun also emits electromagnetic radiation across a variety of wavelengths including radio, infrared, visible light, ultraviolet, and X-rays. Changes in the intensity of these emissions result in the variety of effects known as space weather events, including:

- Sunspots, which can lead to increased emission of solar wind. A geomagnetic storm results, which in mild cases leads to the aurorae borealis and australis, and in more severe cases can overload terrestrial power grids and cause blackouts.

- Coronal mass ejections, which correlate with increased numbers of electrically charged particles being ejected into the solar wind, and which have effects similar to those of sunspots.

- Coronal holes, which also cause increased solar wind activity.

- Solar flares, which result in high-concentration bursts of radiation.

Outside of the aurorae, space weather effects are generally not visible to the naked eye. For the most part, the Earth's natural magnetic field protects the planet from the general solar and space radiation environment. However, when space weather events occur, they can have deleterious impacts on spacecraft operations that operators need to be aware of. These include:

- Higher-than-normal levels of charged particles, which might degrade satellite components and equipment;

- Interference with electrical signals, including those of high-frequency and ultra-high-frequency communications satellites and global navigation satellite systems (GNSS);

- Interference with radar and/or space tracking systems looking in sunward or poleward directions;

- Increased drag for satellites operating in low Earth orbit; and

- The potential for increased radiation exposure for humans in orbit.

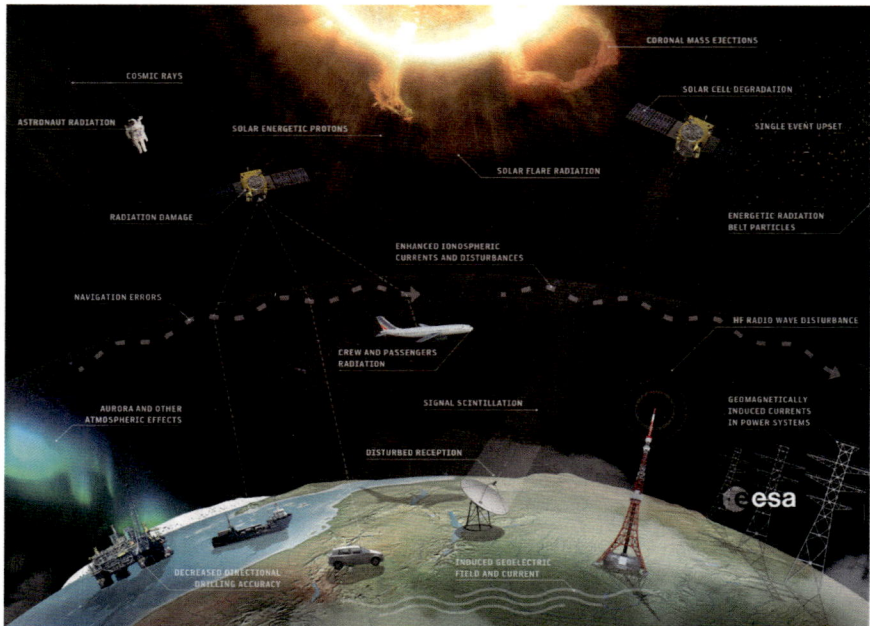

Figure 3.11 | Graphic depicting the effects of space weather on space and terrestrial infrastructure. *Credit: ESA*

Strong space weather events can also impact vulnerable systems on Earth's surface, including electrical power grids and aviation systems.

Space weather is typically correlated with an eleven-year cycle of solar activity, with peaks and troughs termed solar maximum and solar minimum respectively, although notable events can occur at any point in the cycle. Government agencies, such as the National Oceanic and Atmospheric Administration's Space Weather Prediction Center (NOAA SWPC) or the Japan National Institute of Information and Communication Technology, provide space weather forecast services, including offering watches, warnings, and alerts. Depending on the type of space weather event, warnings, watches, and alerts can be issued anywhere from ten minutes to seventy-two hours of advance notice. NOAA SWPC rates space weather events using the Space Weather Scales to describe their expected severity and possible effects on people and systems. Operators and other interested parties can subscribe to the forecast service via their national center or the International Space Environment Service.

Case Study: Starlink Satellite Loss Due to Space Weather

As an example, in early February 2022, 38 of 49 Starlink satellites launched by SpaceX were lost due to unexpectedly accelerated atmospheric decay. The satellites are normally launched into a very low orbit and then after checkout are raised up to their higher operational orbit. However, an unexpected space weather event distended the atmosphere and caused much higher atmospheric drag than usual, which brought the satellites down before they could raise their orbits. As of May 2024, SWPC and other space weather services do not issue alerts or warnings that provide adequate data for satellite users looking at atmospheric drag and related applications.

Satellite Anomaly Recognition, Response, and Recovery

Anomalies in spacecraft operations come in many forms and result from a variety of causes, but are generally described as off-nominal behavior of an individual unit, a subsystem, or the system as a whole. Exact causes of anomalies can cover a broad range of sources, such as the space environment (e.g., high-energy particles from coronal mass ejections, micrometeoroid strikes, spacecraft charging), poor design (e.g., thermal runaway caused by insufficient thermal insulation, divide-by-zero cases within flight software), faulty parts or manufacturing techniques (e.g., debris in bearing races, switch failure), and even procedural or human error during operations (e.g., incorrect sequence of steps for unit power-on, accidentally transmitting unintended commands, unintentional ground- or space-based radiofrequency interference).

At one end of the spectrum, an anomaly may be benign—to the extent that it goes unnoticed for days, weeks, months, or even years. At the other extreme, an anomaly may end a mission. An unexpected or unexplained anomaly may also have military or security implications, especially if it occurs against a backdrop of heightened geopolitical tensions. Properly and thoroughly preparing for, responding to, and learning from anomalies can make the difference between exceeding life expectancies for a mission and experiencing a potentially avoidable mission-ending event.

ANOMALY RECOGNITION

Several steps can be taken to improve an operator's ability to quickly detect anomalies during spacecraft operations. The most important element is having useful, accurate telemetry. All telemetry access points need a clear definition of nominal and off-nominal states or operating ranges. Defining nominal operating ranges generally takes several iterations: the first is the predicted range from unit designers, the second is based on unit test and integration data, and the third is based on initial characterization data.

As insight into the inner workings of a system is only as good as the data available, telemetry format composition should not be overlooked. Not all parameters should be telemetered at the same rate. For example, power failure signatures have very short durations (milliseconds), while thermal signatures generally take time to manifest (seconds to tens of seconds, if not longer). Therefore, power-related data should be telemetered more frequently than thermistor data.

Software components are inherently susceptible to single-event effects (SEEs) caused by energetic particles in the space environment. There is ample literature available on SEEs and methods for designing to account for and respond to them. As a starting point, integrating an error detection and correction (EDAC) capability will help reduce the impact of single-event upsets (SEUs), a type of SEE, but will not fully eliminate the risk of SEUs affecting system performance. Establishing a mechanism to routinely monitor and correct the overall state of data in on-board memory can help catch and correct SEE-related issues before they manifest as errors. In addition, telemetering the status of autonomous corrective actions (quantity, date/ time, location in memory) can provide great insight into the space environment encountered as well as the health of a memory unit itself. For example, repeated attempts to correct the same memory address can provide an indication of a failed or stuck bit.

ANOMALY RESPONSE

Prior to launch, operational procedures should be written, tested, and trained on in order for operators to be adequately prepared to not only perform daily operations but also respond to in-space failures. When developing operational procedures for anomaly response, it is helpful to define strategic decision points in the flow of steps. Consider which steps operators are authorized to execute without supervisory authority and which steps require stakeholder direction (corporate/government/customer) to perform. In defining decision points, also consider what information is necessary to choose the path forward and clearly articulate this information in objective terms. In addition, modularity in procedure design can be useful, as can expected entry/exit states and anticipated duration for the execution of each module.

For LEO systems, if manual intervention is required to respond to an anomalous condition, it must take place during one of the brief in-view periods; therefore, planning quick and concise steps with clear break-points is vital. Prior to the vehicle going out of view, it must be configured to a safe state—a state in which there is little to no risk of further damage or loss of mission until the next in-view period. Similarly, upcoming orbital events in all regimes must be considered. For response to a power system anomaly, for instance, it is important to have heightened awareness of an upcoming eclipse for which the system must be properly charged and configured. If a sufficient state of charge is not possible, a typical response would be to power off noncritical units to allow for safe transit during the eclipse period.

Once all of the above factors have been considered, a system has been built and launched, and in-space operations are underway, failures will inevitably happen. In a perfect world, all failure scenarios have been well-thought-out and detailed operational procedures established along with appropriate responses. In the real world, however, unforeseen and undocumented failures will happen.

When a failure occurs, anomaly response protocol takes effect. The first step in the protocol is an immediate response: any operator action or autonomous fault response sequence required to configure the vehicle to a "safe" state. The second step is to initiate a call-in procedure to alert and request assistance and support from management and system or subsystem experts, based on the observed signature. The third step is establishing authority for action: defining who is in charge of response and recovery actions, which may be the operational crew, factory experts, the owner of the system, or others. The final step is communicating the impact of the anomaly: determining what the immediate effect is on the mission, the duration of the projected outage/impact, and who needs to be informed.

Once a vehicle has been "safed" (configured in a known state it can stay in more or less indefinitely without concern, disregarding a second, unrelated anomaly), operators can begin compiling information about the failure while waiting for system experts to arrive. Useful information includes a detailed timeline of events leading up to the anomaly, detailed state of all systems on the vehicle before and after the fault, and a timeline of upcoming events such as out-of-view periods, eclipses, or conjunctions.

Commercial satellite servicing capabilities are emerging that may be able to add more information from on-orbit sources to help diagnose and understand anomalies. A growing number of companies are offering satellite inspection services, either using retasked Earth imaging satellites or dedicated vehicles, which may be able to provide more data than is possible to gather through telemetry or ground-based imaging. Some of these services may include rendezvous and proximity operations, which will require close coordination between the servicer and client. Adding spacecraft fiducial markers to a satellite before launch can help de-risk such activities.

ANOMALY RECOVERY AND ANALYSIS

An anomaly response team should consist of general vehicle system engineers familiar with the detailed workings of the system as a whole, subsystem and unit specialists educated on specific hardware and software intricacies of the various units, and representatives of the stakeholders or customers. While all satellite operations groups have their own processes for anomaly response, recovery, and investigation, an anomaly recovery usually begins with vehicle system engineers piecing together details of the scenario and working with individual subsystem specialists to identify abnormal behavior in all aspects of a system, both prior to and following the fault. Due to the complexity of space systems and a wide variety of potential causes, many times a specific root cause cannot be attributed on the day of an anomaly. Rather, suspect units are isolated and kept offline until further investigation can take place. In cases where redundant units are available, full operations may be re-established by performing a controlled swap to a redundant unit, if not already performed by onboard fault management.

In general, there are two main severities of anomalies: critical and payload-related. Critical health and safety anomalies affect communications, power, and thermal or attitude control subsystems, and payload-related anomalies may affect execution of the intended mission but do not necessarily affect the ability of the vehicle to control its subsystems. For vehicle health and safety anomalies, autonomous fault management response should be designed and tested to quickly establish safe control of the affected systems. In these cases, the anomaly response team should focus initial efforts on confirming that autonomous commanding successfully identified the fault, executed the proper response sequence, and isolated the suspect unit/s. For noncritical but mission-impacting anomalies, the anomaly team should focus efforts on isolating the fault and investigating the best path forward to re-establishing mission throughput, perhaps on redundant units or in a degraded state if redundant units are not available.

At a point in the anomaly response and recovery process, a full root cause analysis should take place. However, in practice, it is very rarely possible to determine a single definitive root cause. More often, the diagram and paths are narrowed to several "probable root causes" and several "unlikely root causes," and the remainder are "exonerated." Due to the challenges associated with remotely identifying component-level failures from hundreds to millions of kilometers away with limited insight, many root cause investigations remain open, documented with probable but not definitive causes.

FISHBONE DIAGRAMS

Fishbone diagrams are widely recognized as an essential tool to provide a clear and concise way to visually track investigations that have a multitude of potential root causes. "Bones" on the fishbone diagram typically include, at minimum:

- Environmental causes (e.g., space weather, debris, etc.),

- Design/parts/manufacturing causes (down to each piece-part within the failure path), and

- Human/operator causes. As aspects are vetted and eliminated, bones on the fishbone chart can be exonerated. The goal of a deep-dive root cause analysis is to narrow a fishbone diagram down to a single bone that can be deemed the "determined root cause."

Regardless of the absolute determination of the root cause, lessons are always learned from anomalies, lessons which can be applied to the current mission as well as others in a constellation and even across the industry. For example, failure of the bearings in a reaction wheel on a LEO vehicle due to lubrication breakdown may provide early warning of potential trouble with control moment gyros built by the same vendor and used on a different vehicle in a different orbital regime. Therefore, documenting, cataloging, and maintaining failure information is paramount to the success of any space program, as is sharing lessons learned within the space operations community.

Ultimately, it is important to accept that anomalies will happen over the lifetime of a space vehicle. Being adequately prepared before anomalies occur and applying lessons learned afterwards can drastically reduce the impacts to mission success.

END-OF-LIFE

As satellites reach their end of life and cessation of operations, it is important for satellite operators to dispose of satellites properly. Highly used and important regions of orbit are already congested, in large part due to satellites or rocket stages that have been left in those active regions. Increasingly, there are national regulatory obligations, contractual obligations, guarantees, and other responsibilities that need to be met during the end-of-life phase of a space mission.

Post-Mission Disposal

It is important to properly dispose of satellites and launch vehicles at the end of . usable life. Satellites that are not properly disposed of have a chance of interfering with operating satellites and possibly generating additional debris in orbits that are useful and commonly used. To minimize this risk, the Inter-Agency Space Debris Coordination Committee (IADC), an international governmental forum of experts, has created guidelines for mission developers to use when planning proper disposal of spacecraft. In addition, ISO has also developed space systems disposal standards, and several groups of private sector entities have also pledged to adhere to additional practices that go beyond the current international standards.

Launch Vehicle and Satellite Passivation

To minimize the risk of satellites creating debris from accidental breakups after the completion of mission operations, all the onboard sources of stored energy of a spacecraft or launch-vehicle orbital stage, such as residual propellants, batteries, high-pressure vessels, self-destructive devices, and flywheels and momentum wheels, should be depleted or safed when they are no longer required for mission operations or post-mission disposal. This is called passivation.

The importance of designing for proper passivation has been demonstrated by the more than forty ullage motors flown on the Russian Proton Block DM upper stage that have broken up in orbit. The ullage motors, first deployed in the 1980s, are small auxiliary engines that provide the stage with three-axis control during coast and are routinely ejected when the Block DM stage ignites for the final time. Depending on the mission profile, the ullage motors may carry up to 40 kilograms of unused propellant. Over time, solar heating and other factors have caused dozens of the motors to explode, releasing debris into orbit. Russia has made design changes to prevent accidental explosion of the engines on new Block DM models, but some launches continue to eject the units.

Entity		Document
IADC		IADC-02-01, Rev 3 (June 2021)
ISO		ISO 23112 (2022), ISO 24113: (2023)
USA		U.S. Government Orbital Debris Mitigation Standard Practices (November 2019)
	NASA	NPR 8715.6B (February 2017), NASA-STD-8719.14 (November 2021)
	DoD	DoD Directive 3100.10 (August 2022)
	FAA	Title 14, Code of Federal Regulation (CFR) Part 415.39
	FCC	Space Innovation IB Docket No. 22-271 Mitigation of Orbital Debris in the New Space Age IB Docket No. 18-313
JAXA		JAXA Space Debris Mitigation Standard JMR-003E (April 2023)
CNES		Technical Regulation (TR) of the French Space Operations Act (FSOA)
ESA		ESA Space Debris Mitigation Requirements ESSB-ST-U-007
Roscosmos		GOST R 52925-2018
China		State Administration of Science,. Technology and Industry for National Defense, notice on promoting the orderly development of microsatellites and strengthening safety management

Table 3.3 | International orbital debris limitation documents.

Passivation should occur as soon as the process can be undertaken without posing unacceptable risk to the satellite payload. Guidelines include the following:

- Residual propellants and other fluids, such as pressurants, should be depleted as thoroughly as possible, either by depletion burns or venting, to prevent accidental breakups caused by overpressurization or chemical reaction.

- Batteries should be designed and manufactured, both structurally and electrically, to prevent breakups. Pressure increase in battery cells and assemblies can be prevented by mechanical measures unless these measures cause an excessive reduction of mission assurance. At the end of operations, battery charging lines should be deactivated.

- High-pressure vessels should be vented to a level guaranteeing that no breakups can occur. Leak-before-burst designs are beneficial but are not sufficient to meet all passivation recommendations of propulsion and pressurization systems. Heat pipes may be left pressurized if the probability of rupture can be demonstrated to be very low.

- Self-destruct systems should be designed to not cause unintentional destruction due to inadvertent commands, thermal heating, or RF interference.

- Power to flywheels and momentum wheels should be terminated during the disposal phase.

- Other forms of stored energy should be assessed and adequate mitigation measures should be applied.

- All communications, including telemetry and other forms of RF transmission, should be turned off or disabled.

Geosynchronous Region Disposal

The geosynchronous region is a special area of Earth orbit. This is defined as 200 kilometers above and below the geostationary altitude of 35,786 kilometers and 15 degrees north and south of the Equator. Maintaining a spacecraft in GEO in the geosynchronous region requires expenditure of fuel over time to maintain a fixed position over a point on the surface of the Earth. GEO satellites are disposed of by maneuvering the spacecraft further out into space, away from the protected geosynchronous region. However, the decision on when to retire a GEO satellite can be a difficult tradeoff.

GEO satellites often face depletion of fuel before other satellite subsystems reach end-of-life. Therefore, operators often must make the difficult decision to retire a satellite that is generating tens of millions of dollars annually, when the only thing wrong with it is its low fuel. The lifespan tradeoff is made more difficult because satellite operators, using newly available low-cost tracking user terminals, can choose to conduct operations from an inclined orbit. In an inclined orbit, fuel is expended at a much reduced rate as the satellite is allowed to drift within a certain region of space, allowing it to continue to be useful. However, there is a risk that other satellite subsystems, operating beyond their design life, may fail during inclined operations, leaving the satellite in an orbit that risks contaminating the protected geosynchronous region.

The IADC recommends fulfilling the two following conditions at the end of the disposal phase to describe an orbit that remains above the geosynchronous protected region:

1. A minimum increase in perigee altitude of:
   235 km + (1000 × CR × A/m) where
   CR is the solar radiation pressure coefficient,
   A/m is the aspect area to dry mass ratio (m²kg⁻¹), and
   235 km is the sum of the upper altitude of the GSO protected region (200 km) and the maximum descent of a re-orbited spacecraft due to luni-solar & geopotential perturbations (35 km)

2. An eccentricity less than or equal to 0.003

To minimize the chance of debris creation, a propulsion system should not be separated from a GEO spacecraft. In the event that there are unavoidable reasons that require separation, the propulsion system should be designed to be left in an orbit that is, and will remain, outside of the protected geosynchronous region. Regardless of whether it is separated or not, a propulsion system should be designed for passivation. In addition, spacecraft operators should design missions to avoid leaving launch vehicle orbital stages in the geosynchronous region. Most GEO operators require that manufacturers design for one more year than required for operation so that the satellite can be moved above the geostationary orbit and allowed to drift away into deep space.

Passing Through LEO Disposal

Some types of launches leave rocket bodies or other fragments in orbits that pass through LEO. Often this is the case with launches to place a GEO satellite into a geostationary transfer orbit (GTO), a navigation satellite in medium Earth orbit (MEO), or a satellite in highly elliptical Molniya orbits. Whenever possible, spacecraft or orbital stages that are terminating their operational phases in orbits that pass through the LEO region, or have the potential to interfere with the LEO region, should be de-orbited (direct re-entry is preferred), or, where appropriate, maneuvered into an orbit with a reduced lifetime. Retrieval is also a disposal option.

According to the IADC, a spacecraft or orbital stage should be left in an orbit in which, using an accepted nominal projection for solar activity, atmospheric drag will limit the orbital lifetime after completion of operations to twenty-five years. If a spacecraft or orbital stage is to be disposed of by re-entry into the atmosphere, debris that survives to reach the surface of the Earth should not pose an undue risk to people or property. To minimize the risk of debris surviving re-entry, it is advisable to design a satellite in a manner that results in complete vaporization during re-entry. If that is not possible and there is a greater than 1-in-10,000 chance of causing a fatality, it is necessary to perform a controlled re-entry that deposits surviving debris into uninhabited regions, such as broad ocean areas. In addition, ground environmental pollution, caused by radioactive substances, toxic substances, or any other environmental pollutants resulting from onboard articles, should be prevented or minimized in order to be accepted as permissible.

> If debris is expected to survive re-entry and cause an unacceptable risk of casualties, it is necessary for mission planners to conduct a controlled re-entry that will spread debris over uninhabited areas of the Earth's surface.

In the event of a controlled re-entry of a spacecraft or orbital stage, the operator of the system should inform the relevant air traffic and maritime traffic authorities of the re-entry time and trajectory and the associated ground area.

Atmospheric Re-Entry and Risk Assessment

Spacecraft designers must consider what will happen to a spacecraft at the end of its lifespan. For satellites operating in LEO, it is likely that atmospheric drag will eventually cause a spacecraft to re-enter Earth's atmosphere. As satellites re-enter, they disintegrate, but some debris may survive the heat of re-entry and could impact the ground and cause casualties. Unfortunately, it is very difficult to predict specifically where debris will impact as the density of the Earth's atmosphere is constantly changing. It is recommended that satellite operators design spacecraft that will completely burn up during re-entry.

If debris is expected to survive re-entry and cause an unacceptable risk of casualties, it is necessary for mission planners to conduct a controlled re-entry that will spread debris over uninhabited areas of the Earth's surface.

RE-ENTRY

During re-entry, friction and compression generate immense heat as a satellite traveling more than 29,000 kilometers per hour enters the atmosphere. That tremendous heat can melt and vaporize the entire spacecraft. However, if a satellite component's melting temperature is not reached during re-entry then that object can survive re-entry and impact the ground. In addition to heat and pressure, a spacecraft experiences immense loads as it decelerates. These loads, which can exceed 10 Gs. or ten times the acceleration of gravity at the Earth's surface, coupled with the immense heat, cause a spacecraft's structure to break apart. The broken-up components will continue to decelerate and, depending on the density of the atmosphere in the region of re-entry, may reach a low ground speed, virtually falling straight down from the sky. The broken-up spacecraft should impact the ground at relatively low speeds, but it still presents a hazard to people and property on the ground and a satellite operator will be liable for damages caused by the debris. For more information, see Chapter One: International Liability and Chapter Two: National Laws.

Predicting the exact area where debris will impact from an uncontrolled re-entry is difficult because drag on the object is directly proportional to atmospheric density, and the density of the atmosphere varies greatly at high altitudes and is affected—dramatically even—by solar activity. It is possible to predict the time a re-entry will begin within a 10 percent margin of the actual time. However, even one minute of error in time corresponds to a difference of hundreds of kilometers on the ground because of the great speeds of re-entering objects.

About 10 to 40 percent of a satellite's mass may survive re-entry, depending on the size, shape, weight, and material composition. The area of highest probability where it will strike the surface of the Earth and potentially produce a debris field is called a footprint. It is possible to predict the size of a footprint but very difficult to determine specifically where the debris footprint will be located on the Earth's surface. The size of the footprint is determined by estimating the breakup altitude of the satellite or space hardware and then modeling the mass and aerodynamic properties of surviving debris. Footprint lengths vary in size from

Figure 3.12 | The final re-entry breakup process, showing that debris surviving re-entry would fall through airspace potentially occupied by aircraft and could spread over a long, narrow path. *Credit: The Aerospace Corporation*

approximately 185 kilometers to 2,000 kilometers, depending on the complexity and characteristics of the object. The width of a footprint can be affected by winds, with the greatest uncertainty affecting the lightest objects. A 20- to 40-kilometer footprint width is typical.

RE-ENTRY THREAT CONSIDERATIONS
While the impact threat to human life and property from the atmospheric re-entry of space debris is serious, only one person has ever claimed to have been struck by falling space debris, and that person was hit by a lightweight object and was not injured. In April 2024, a piece of debris from the International Space Station crashed through the roof of an occupied house in Florida without injuring anyone. Over the last fifty years, a significant amount of material is known to have survived re-entry, but no casualties from the debris have been reported.

CALCULATING RE-ENTRY RISK
There is no internationally accepted standard of "acceptable safety risk" for re-entry. The UN COPUOS space debris mitigation guidelines leave the delineation of acceptable risk to national authorities. The IADC identifies two guidelines to follow. First, to minimize the accumulation of orbital debris, it recommends that satellites in LEO that have completed their missions should be left in an orbit that will result in re-entry within twenty-five years. While compliance with this recommendation is not perfect, most major spacefaring nations support the "25-year rule" and

are taking steps to improve compliance. Studies done by the European Space Agency indicate that since 2010, successful compliance with the 25-year rule has gone from 10 percent to 55 percent for payloads launched into LEO and from 20 percent to 80 percent for upper stages. Over the same time period, successful compliance for GEO payloads has gone from 60 percent to nearly 90 percent.

In addition to the 25-year rule, the IADC recommends that if a satellite has a 1-in-10,000 chance of surviving re-entry and causing a casualty (either a fatality or serious injury requiring hospitalization), its re-entry must be controlled. For a piece of debris that survives atmospheric re-entry, the debris casualty area is the average debris cross-sectional area plus a factor for the cross-section of a standing individual. The total debris casualty area for a re-entry event is the sum of the debris casualty areas for all debris pieces that survive atmospheric re-entry. The total human casualty expectation is equal to the total casualty debris area times the average population density for the particular orbit. A variety of models exist to calculate the likelihood that specific pieces of a satellite will survive re-entry, including NASA's Debris Assessment Software (DAS) or its higher-fidelity Object Re-entry Survival Analysis Tool (ORSAT) and ESA's Meteoroid and Space Debris Terrestrial Environment Reference (MASTER) and Debris Risk Assessment and Mitigation Analysis (DRAMA).

RE-ENTRY PREDICTIONS

Spacecraft re-entries are tracked by space surveillance systems around the globe. These ground- and space-based radars and telescopes can be used along with models of the Earth's atmosphere to determine when an object is likely to re-enter the atmosphere and the potential path of re-entry, although not specific ground impact locations. One major source of these re-entry predictions is the U.S. military, which shares information on upcoming re-entries publicly on spacetrak.org. Tracking and Impact Prediction messages are issued at T–4 days, T–3 days, T–2 days, T–1 day, T–12 hours, T–6 hours, and T–2 hours. Re-entry predictions must be continually updated as a satellite approaches the atmosphere.

Even predictions made within a few hours of re-entry may project a debris footprint that is incorrect by hundreds to thousands of kilometers. Therefore, even if a significant amount of debris is expected to survive re-entry, it is not logistically possible to evacuate areas debris might impact. The predictions are mainly used to alert governments about potential impact events that may threaten populated areas so that first responders and other services can prepare accordingly.

PLANNING A CONTROLLED RE-ENTRY

If significant portions of a satellite are expected to survive re-entry and violate the 1-in-10,000 chance casualty threshold, it is important for a satellite designer to plan a controlled re-entry that will scatter any remaining debris over an unpopulated part of the ocean. A controlled re-entry requires a satellite maneuvering strategy that avoids possible collision with space debris or other satellites. Adequate fuel must be left in a satellite's tanks to perform the final orbit-changing burns. Ground support teams must be available to coordinate, perform, and monitor the final satellite maneuvers.

EMERGING ISSUES

The rapid increase in the number and types of activities taking place in the space environment, driven in part by novel private sector activities, is raising a number of issues and challenges for which responsible operations practices and governance frameworks are not yet set. These issues include areas where space activity may have unanticipated impacts on the environment, operations where safety practices need to be established and codified, and development of practices for new types of space activity.

The rapid increase in the number and types of activities taking place in the space environment, driven in part by novel private sector activities, is raising a number of issues and challenges for which responsible operations practices and governance frameworks are not yet set.

Orbital Carrying Capacity

As governments and companies carry out the planning, deployment, and operations of multiple very large satellite constellations consisting of thousands to tens of thousands of individual satellites in the same or similar orbits, concern has been raised whether there is a limit to the capacity of specific orbital regimes. To better understand this concern, research is being conducted within the space debris and orbital dynamics scientific community to define and assess a concept known as "orbital carrying capacity," which can be loosely described as a quantification of the amount of activity specific orbital regions can sustain. Carrying capacity is concerned with both the ability to effectively manage orbital regimes for current operators and the ability to maintain access for future activities and operators. There are a range of research approaches and concepts for defining and assessing orbital carrying capacity, including long-term environmental perspectives, near-term space safety perspectives, behavioral approaches, and approaches looking at economic incentives.

As of May 2024, orbital carrying capacity concepts have not reached a level of research maturity where they are operationally relevant to regulatory or licensing considerations, nor is it clear how the regulatory process might be best equipped to utilize carrying capacity. However, continuing research into orbital carrying capacity may ultimately help establish a more holistic approach to space environmental management.

Potential Atmospheric Effects from Launches and Re-Entries

As the rate of space launch and the numbers of satellites re-entering the Earth's atmosphere both continue to increase, researchers have begun to study the potential effects those increased activities might have on the atmosphere. Early results from this research, while preliminary, indicate that there is potential for the accumulation of these space activities to have deleterious or negative effects on atmospheric chemistry and composition, and on climate change. NOAA research has found that these potential effects include the deposition of metallic particles in the upper atmosphere as a result of satellites and rocket bodies burning up during re-entry as well as the deposition of microparticles containing black carbon soot, nitrous oxides, and alumina in the emissions from rocket launches, both of which could contribute to climate change and ozone layer depletion.

Research efforts to describe and quantify these potential impacts are in initial stages, and there is a need to both increase this research and to begin to examine potential mitigation approaches. Mitigation approaches might include changes in materials used for satellite manufacturing, revisions to design-for-demise practices, and increased emphasis on clean or "green" fuels. In some cases, as research progresses, there is potential that best practice for mitigation of one space sustainability challenge (e.g., post-mission disposal to minimize space debris) may be in tension with other impacts of space activities on the environment (e.g., atmospheric effects of satellite re-entry).

Sustainable Practices for Cislunar Space and Lunar Surface Operations

Space activities are increasingly expanding into cislunar space and onto the lunar surface. However, operational best practices for this area of activity are not as well defined as for operations in GEO and LEO. Many of the practices that have been developed for operations in GEO and below will not suffice or even apply to the different physical and operational environment in the cislunar regime. There are currently no established consensus practices for mitigating the creation of debris in lunar orbit and on the lunar surface nor for post-mission disposal of spacecraft in cislunar space. Furthermore, space situational awareness capabilities to monitor activities in the cislunar domain are currently very limited. On the lunar surface itself, dealing with the lunar dust disturbed by lunar activities will be a shared challenge faced by all lunar operators. There will also be a need to develop interoperability in key infrastructure in the interests of safety. Running through all of these challenges is the need for coordination and information exchange related to lunar activities. Currently no formal coordination mechanisms for lunar activities currently exist.

This lack of lunar coordination mechanisms could lead to misunderstandings and safety challenges and even the potential for dangerous misunderstandings. As the tempo of activities in the cislunar domain increases there is need to develop specific operations practices for that domain, and refine those practices based on lessons learned from shared operational experience. There is also a need to develop formal coordination forums at the international level. Informal groups in the lunar community such as the Global Expert Group on Sustainable Lunar Activities have begun to develop initial suggestions for such practices and coordination mechanisms.

In-Space Servicing, Assembly, and Manufacturing

In-Space Servicing, Assembly, and Manufacturing (ISAM) capabilities represent a significant change in how space activities are conducted and could in turn lead to many advances in space safety and sustainability. However, while there has been some initial work done by groups such as CONFERS to develop recommended practices and standards for commercial ISAM activities, there is still much work to do and many aspects of ISAM are still in a nascent state. The inherently dual-use nature of ISAM capabilities pose special challenges for ensuring the transparency of intentions and activities to help prevent misperceptions and mistakes that can increase tensions. Use of commercial ISAM capabilities to support national security space missions adds to this challenge.

The novelty of ISAM capabilities also remains a challenge for international and national oversight frameworks, including spectrum allocation and coordination, many of which were developed without considerations for ISAM. As commercial ISAM capabilities continue to evolve and mature, there will need to be a similar evolution of the oversight frameworks to include consideration of ISAM capabilities.

Commercial Human Spaceflight Safety

Historically, conducted solely by governmental operators, human spaceflight activities (both suborbital and orbital) are increasingly being conducted by commercial operators. These commercially and privately operated flights now include both solely commercial missions and missions conducted by companies on behalf of governments, including flying government personnel as commercial spaceflight participants. Commercial human spaceflight is still generally considered an experimental activity, in which participants take part under an informed consent regime. However, as commercial human spaceflight becomes more common, there will be a need to document and codify safety practices, including lessons learned. Ultimately the informed consent regime will need to be replaced with more formal and systematic certification, inspection, and regulation based safety practices. National jurisdictions which currently have learning periods or regulatory moratoria in place regarding commercial human spaceflight will eventually need to lift those measures and put into place effective and appropriate regulatory frameworks.

CHAPTER THREE ADDITIONAL RESOURCES

See below for additional resources, links, and documents referenced in Chapter Three.

ASCEND Satellite Orbital Safety Best Practices: https://www.ascend.events/outcomes/satellite-orbital-safety-best-practices-by-iridium-oneweb-spacex-aiaa/?_gl=1*1phohy*_ga*MjA0MTQyNDMwMi4xNzE0NTcwMTE5*_ga_BFMKMMYM72*MTcxOTI3NDk2NC45LjAuMTcxOTI3NDk2NC42MC4wLjA.

CONFERS Recommended Design and Operational Practices: https://www.satelliteconfers.org/wp-content/uploads/2021/11/CONFERS_Operating_Practices_Revised-Oct-21.pdf

CONFERS: https://satelliteconfers.org/

Consultative Committee for Space Data Systems: https://public.ccsds.org/

DoD Directive 3100.10 (August 2022): https://technology.esa.int/upload/media/ESA-Space-Debris-Mitigation-Requirements-ESSB-ST-U-007-Issue1.pdf

ESA DRAMA Python Package: https://sdup.esoc.esa.int/drama/python_package_docs/

ESA MASTER Software Tool: https://www.esa.int/ESA_Multimedia/Images/2013/04/ESA_s_MASTER_software_tool

ESA Reentry and Collision Avoidance: https://www.esa.int/Space_Safety/Space_Debris/Reentry_and_collision_avoidance

ESA Space Debris Conference– An overview of Design for Demise Technologies: https://conference.sdo.esoc.esa.int/proceedings/sdc8/paper/130

ESA Space Debris Mitigation Requirements: https://technology.esa.int/upload/media/ESA-Space-Debris-Mitigation-Requirements-ESSB-ST-U-007-Issue1.pdf

ESA Space Environment Report: https://www.sdo.esoc.esa.int/environment_report/Space_Environment_Report_latest.pdf

ESA Zero Debris Charter: https://esoc.esa.int/sites/default/files/Zero_Debris_Charter_EN.pdf

EU Space Surveillance and Tracking: https://www.eusst.eu/

European Space Security and Innovation (ESSI): https://www.essi.org/#top

FAA Commercial Space Launch and Reentry Financial Responsibility: https://www.faa.gov/space/licenses/financial_responsibility

FAA Environmental Assessment for SpaceX Starship/Super Heavy at Boca Chica: https://www.faa.gov/sites/faa.gov/files/2022-06/PEA_for_SpaceX_Starship_Super_Heavy_at_Boca_Chica_FINAL.pdf

FAA Flight Safety Analysis Handbook: https://www.faa.gov/about/office_org/headquarters_offices/ast/media/Flight_Safety_Analysis_Handbook_final_9_2011v1.pdf

French Space Operations Act: https://www.legifrance.gouv.fr/loda/id/JORFTEXT000024095828/

Global Expert Group on Sustainable Lunar Activities: https://moonvillageassociation.org/global-expert-group-on-sustainable-lunar-activities-gegsla/#

GSOA Code of Conduct on Space Sustainability: https://gsoasatellite.com/reports_and_studies/global-satellite-operators-association-releases-code-of-conduct-on-space-sustainability/

International Space Environment Service (ISES): http://www.spaceweather.org/index.jsp

ISO 23312 – 2022 Space systems — Detailed space debris mitigation requirements for spacecraft: https://www.iso.org/standard/75221.html

ISO 24113 – 2023 Space systems — Space debris mitigation requirements: https://www.iso.org/standard/83494.html

ISO 24330 Space Systems - Rendezvous And Proximity Operations (RPO) and On Orbit Servicing (OOS) - Programmatic Principles and Practices: https://webstore.ansi.org/standards/iso/iso2433020221

ISO 27875: https://www.iso.org/obp/ui/en/#iso:sttd:iso:27875:ed-2:v1:en

Japan National Institute of Information and Communication Technology: https://swc.nict.go.jp/

JAXA Space Debris Mitigation Guidelines: https://sma.jaxa.jp/TechDoc/Docs/E_JAXA-JMR-003E.pdf

MITRE ATT&CK: https://attack.mitre.org/

NASA Object Re-entry Survival Analysis Tool (ORSAT): https://orbitaldebris.jsc.nasa.gov/reentry/orsat.html

NASA OCE Report: https://www.nasa.gov/wp-content/uploads/2023/07/oce-51.pdf?emrc=c0a365?emrc=c0a365

NASA Orbital Debris Program Office – Debris Assessment Software: https://www.orbitaldebris.jsc.nasa.gov/mitigation/debris-assessment-software.html

NASA Process for Limiting Orbital Debris : https://standards.nasa.gov/sites/default/files/standards/NASA/C/0/nasa-std-871914c.pdf

NASA Safety Standard for Probabilistic Risk Assessment: https://nodis3.gsfc.nasa.gov/displayDir.cfm?t=NPR&c=8715&s=6B

NASA Space Security Best Practices Guide: https://swehb.nasa.gov/display/SWEHBVD/7.22++Space+Security%3A+Best+Practices+Guide

NASA–NanoRacks Cubesat Deployment: https://www.nasa.gov/image-article/set-of-nanoracks-cubesats-deployed-from-international-space-station/

NOAA Research – Exotic Metal Particles in the Upper Atmosphere Linked to Rockets and Satellites: https://research.noaa.gov/2023/10/16/noaa-scientists-link-exotic-metal-particles-in-the-upper-atmosphere-to-rockets-satellites/

NOAA Space Weather Prediction Center: https://www.swpc.noaa.gov/

NOAA Space Weather Scales: https://www.swpc.noaa.gov/noaa-scales-explanation

Space Data Association: https://www.space-data.org/sda/

Space Innovation IB Docket No. 22-271 Mitigation of Orbital Debris in the New Space Age IB Docket No. 18-313: https://docs.fcc.gov/public/attachments/FCC-22-74A3.pdf

Space Safety Coalition Best Practices: https://spacesafety.org/wp-content/uploads/2024/05/SSC_Best_Practices_for_Space_Operations_Sustainability_v2.38.pdf

Space Safety Institute (SSI): https://spacesafety.org/

Space Sustainability Rating: https://spacesustainabilityrating.org/

Space-Track.org: http://spacetrak.org/

SWF Counterspace Report: https://swfound.org/counterspace

Title 14, Code of Federal Regulation (CFR) Part 415.39: https://www.ecfr.gov/current/title-14/chapter-III/subchapter-C/part-415

U.S. Government Orbital Debris Mitigation Standard Practices: https://orbitaldebris.jsc.nasa.gov/library/usg_orbital_debris_mitigation_standard_practices_november_2019.pdf

U.S. Space Force Combined Space Operations Center: https://www.vandenberg.spaceforce.mil/Units/CSpOC-DEL-5/

United Arab Emirates Space Agency: https://space.gov.ae

RESPONSIBLE OPERATIONS IN SPACE

Annex

Relevant Organizations and Fora

INTRODUCTION

There are a multitude of fora where various actors seek to engage in space policy-related decision-making, advocacy, and expertise. Below is a list of organizations new actors should be aware of when conducting space activities. The list below is not exhaustive, but it encapsulates an overview of significant organizations that new actors should understand and consider engaging.

There are three broad categories of organizations:

» Intergovernmental Organizations, Agencies, and Fora;

» Trade Associations and Industry Groups; and

» Scientific and Technical Organizations.

Each category is defined based primarily on who comprises their membership and the interests they represent and may have exceptions or nuances not captured in this text.

INTERGOVERNMENTAL ORGANIZATIONS, AGENCIES, AND OTHER FORA

The selection of organizations below shows some of the international groups that focus, in whole or in part, on sustaining the use of outer space for all. These organizations are characterized by having member states as the primary participants and often deal with norm-building, standard-setting, and rule-making for space activities. Others serve as venues for technical and other international collaboration efforts. This category also includes regional organizations focused on similar activities. Observer states, NGOs, or other groups may also be present for the proceedings of these organizations.

African Space Agency (AfSA)

The African Space Agency (AfSA) was officially established on January 25, 2023. It is composed of member states from the African Union (AU), aiming to promote the continent's space activities, coordination, and research. It is headquartered in Cairo, Egypt, and seeks to promote cooperation between the different space policies of AU members.

Asia-Pacific Regional Space Agency Forum (APRSAF)

The Asia-Pacific Regional Space Agency Forum was formed in 1993 to enhance space activities in the Asia-Pacific region. It is composed of space agencies, governmental bodies, international organizations, private companies, universities, and research institutes from over forty countries. Additionally, APRSAF has five working groups and a workshop group. These additional entities focus on a variety of topics, including satellite navigation, space exploration, and engineering management.

Asia-Pacific Space Cooperation Organization (APSCO)

The Asia-Pacific Space Cooperation Organization is an intergovernmental organization that seeks to promote and strengthen the development of collaborative space programs among its Member States by establishing the basis for cooperation in peaceful applications of space science and technology. It is headquartered in Beijing and has eight member states.

Committee on Earth Observation Satellites (CEOS)

The Committee on Earth Observation Satellites was formed in 1984. It is composed of space agencies and organizations from around the world. CEOS works by coordinating satellite missions and data sharing. It aims to optimize global Earth observation efforts to monitor and address environmental and climate challenges.

European Space Agency (ESA)

The European Space Agency was established on May 30, 1975, and is composed of twenty-two member states from across Europe. ESA missions include scientific exploration, Earth observation, and telecommunications. It aims to explore space, develop satellite technologies, and foster international cooperation between European countries and the world.

The European Union Agency for the Space Programme (EUSPA)

The European Union Agency for the Space Programme (EUSPA) was formed in 2021. It is composed of EU member states and space industry stakeholders. EUSPA works by managing and promoting EU space programs, including Galileo and Copernicus. Its aim is to ensure the efficient use of space technologies for economic growth and security.

Group on Earth Observations (GEO)

Established in 2005, and as of 2024 having 116 member countries, 151 participating organizations, and 19 associates, GEO co-produces user-driven Earth intelligence solutions that inform decisions and accelerate action on global, societal, and environmental challenges, leveraging its unique position as an established intergovernmental body with a strong and inclusive partnership.

ANNEX

Inter-Agency Space Debris Coordination Committee (IADC)

Founded in 1993, the IADC is an international governmental forum of thirteen space agencies focused on the worldwide coordination of activities related to artificial and natural debris in space. The IADC formulated technical space debris mitigation guidelines in 2002, revised them in 2007, and updated them in 2021. The IADC is composed of four working groups: measurements (WG1), environment and database (WG2), protection (WG3), and mitigation (WG4), as well as a Steering Committee.

International Telecommunication Union (ITU)

The ITU is a specialized agency of the UN system and is based in Geneva, Switzerland. It is tasked with facilitating equitable access to the electromagnetic spectrum and orbital resources for satellite services and with promoting the advancement, implementation, and efficient operation of these services. The ITU manages the international frequency coordination process, develops global standards, and maintains the Master International Frequency Register. Every three to four years, the ITU also convenes the World Radiocommunication Conference to revise or adopt the international Radio Regulations—a treaty containing the regulatory, operational, procedural, and technical provisions applicable to radio spectrum and orbital resources. Each country has one vote at the WRC, though many decisions are adopted by consensus.

ITU Radiocommunication Sector (ITU-R)

The International Telecommunication Union's Radiocommunication Sector (ITU-R) was established in 1992. It operates through study groups, conferences, and working parties that develop international standards and regulations for radiocommunication services. It is composed of ITU member states, private sector members, and academic institutions. ITU-R aims to ensure efficient and interference-free use of the radio-frequency spectrum and satellite orbits. Its goals include facilitating global wireless communication, enhancing spectrum management, and promoting the development of new radiocommunication technologies.

ITU Telecommunication Standardization Sector (ITU-T)
The Telecommunication Standardization Sector (ITU-T) is the division of the ITU responsible for coordinating technical standards for telecommunications. It does this through a consensus-based approach with both member states and sector members providing input to the numerous study groups. The purpose of the study groups is to develop "Recommendations" and other technical documents, which become mandatory only when adopted as part of a national law. The World Telecommunication Standardization Assembly (WTSA), which meets every four years, approves the study groups, sets their work programme for the next four-year period, and appoints their chairmen and vice-chairmen.

Though not as relevant for space actors as the ITU-R, the ITU-T has study groups looking at cybersecurity, the Internet of Things (IoT), 5G, and other topics of interest to some companies.

ITU Development Sector (ITU-D)
The International Telecommunication Union's Development Sector (ITU-D) was established in 1992. Its primary aim is to promote equitable and sustainable access to information and communication technologies (ICTs) worldwide. It is composed of ITU member states and sector members. ITU-D focuses on reducing the digital divide, fostering digital inclusion, and supporting developing countries in enhancing their ICT infrastructure and capabilities. It provides technical assistance, policy guidance, and capacity-building programs.

International Organization for Standardization (ISO)
Headquartered in Geneva, Switzerland, the ISO is an independent organization with a membership of 163 national standards bodies. Through its members, it brings together experts to share knowledge and develop voluntary, consensus-based, and market-relevant international standards that support innovation and provide solutions to global challenges. ISO maintains a standing Technical Committee on Aircraft and Space Vehicles (TC20), and subcommittees on Space Data and Information Transfer Systems (SC13) and Space Systems and Operations (SC14).

Intersputnik
Intersputnik was formed on November 15, 1971. It is composed of member countries from Eastern Europe, Asia, Africa, and Latin America. The organization pools satellite resources for affordable communication services, aiming to enhance global telecommunications, especially for developing countries, through cooperation and satellite technology.

ANNEX

United Nations (UN)
Established by the Charter of the United Nations in 1945, the UN is the world's largest and most important international intergovernmental political institution. Its principal organs (the General Assembly, Security Council, Economic and Social Council, Secretariat, and International Court of Justice) work to maintain international peace and security; cooperate in solving international economic, social, cultural, and humanitarian problems; and promote respect for human rights and fundamental freedoms.

United Nations General Assembly (UNGA)
The General Assembly is the UN's main deliberative organ and is composed of all Member States, who each have one vote on all decisions. The General Assembly meets in New York at the UN Headquarters each year in the second half of September. Decisions on important matters (such as peace, security, and new members) require a two-thirds majority, while all other matters require a simple majority. Much of the work of the General Assembly is carried out by its committees and other bodies, two of which concern themselves with matters related to outer space.

First Committee
The First Committee of the UN General Assembly is the Disarmament and International Security Committee, which tasks itself with general disarmament and international security issues that occasionally touch upon issues related to outer space.

Fourth Committee
The Fourth Committee of the UNGA is the Special Political and Decolonization Committee. The yearly report from COPUOS is received by the Fourth Committee, which also creates the mandate for the next year of work by COPUOS.

United Nations Conference on Disarmament (UN CD)
Headquartered in Geneva, Switzerland, as a successor to previous UN-organized committees related to disarmament, the current Conference on Disarmament (CD) was established in 1980 and deals with a number of issues interrelated with disarmament, including as a regular agenda item the prevention of an arms race in outer space (PAROS).

United Nations Office for Outer Space Affairs (UN OOSA)
Headquartered in Vienna, Austria, OOSA is organized under the UN Secretary-General and has two sections: the Committee, Policy, and Legal Affairs section, and the Space Applications Section. OOSA acts as the Secretariat to COPUOS and to its two subcommittees. OOSA's Programme on Space Applications assists developing countries in using space technology for development, providing technical assistance, training, and fellowship programs in remote sensing, satellite communication, satellite meteorology, satellite navigation, space law, and basic space sciences. OOSA is also the keeper of the UN registry of space objects and serves as the secretariat to the ICG. OOSA also manages the

Platform for Space-based Information for Disaster Management and Emergency Response (UN-SPIDER) with offices in Vienna, Austria; Bonn, Germany; and Beijing, China.

United Nations Committee on the Peaceful Uses of Outer Space (UN COPUOS)
Established by a UNGA resolution in 1958, COPUOS is the principal UN committee considering space activities. COPUOS and its two subcommittees meet in Vienna, Austria. The Scientific and Technical Subcommittee (STSC) meets for two weeks, usually in January or February, the Legal Subcommittee (LSC) meets for two weeks sometime in March, April, or May, and the full COPUOS plenary meets in the summer for one and a half weeks. As of 2024, membership (which is only open to states) is 102 and growing, and a diverse number of intergovernmental and non-governmental permanent observers also attend. Reports from COPUOS are sent for approval to the UNGA's Fourth Committee. COPUOS is the body where the principal legal instruments, such as the Outer Space Treaty, were drafted and negotiated. For a more detailed understanding of COPUOS's history and workings, please see the Secure World Foundation publication, *The COPUOS Briefing Book*.

UN Committee of Experts on Global Geospatial Information Management (UN-GGIM)
Established in 2011, UN-GGIM provides a forum for coordination and exchange among member states and international organizations while promoting the development of global geospatial information and its use in addressing global challenges.

World Meteorological Organization (WMO)
Established in 1950, the WMO is a specialized agency of the United Nations dedicated to international cooperation in the areas of meteorology, hydrology, and related applications. The WMO facilitates policy formulation and exchange of data related to these areas and maintains a number of reference standards and datasets. The WMO Space Programme works to coordinate the availability and utilization of space-based data sources and products for weather and climate observation purposes in the WMO's 191 member states.

International Committee on Global Navigation Satellite Systems (ICG)
Established in 2005, the ICG strives to promote voluntary cooperation on matters of mutual interest related to civil satellite-based positioning, navigation, timing (PNG), and value-added services. This includes coordination among providers of GNSS, regional systems, and augmentations to ensure greater compatibility, interoperability, and transparency.

ANNEX

TRADE ASSOCIATIONS AND INDUSTRY GROUPS

This selection of organizations focuses on private sector space organizations. These organizations are often composed of individual companies or conglomerates. These groups often work closely with governments through advocacy and outreach, maintaining a central focus on preserving the outer space environment for societal and economic purposes.

Aerospace Industries Association (AIA)

The Aerospace Industries Association (AIA) was formed in 1919, and is composed of major aerospace and defense companies in the United States. AIA aims to promote the aerospace industry's interests, advance innovation, and support national security and economic growth. It seeks to accomplish these aims through advocacy, policy development, and industry collaboration.

The American Institute of Aeronautics and Astronautics (AIAA)

The American Institute of Aeronautics and Astronautics (AIAA) was formed in 1963. It is composed of aerospace professionals and enthusiasts, operating through conferences, publications, and technical committees. It aims to advance aerospace science, technology, and engineering, foster professional development, and promote the aerospace industry's contributions to society.

Asia-Pacific Satellite Communications Council (APSCC)

The Asia-Pacific Satellite Communications Council is an international nonprofit association representing all sectors of the satellite and/or space-related industry. Its members include satellite manufacturers, launch service providers, satellite service providers and satellite risk management companies, telecom carriers, and broadcasters from Asia, Europe, and North America. APSCC's overall mission is to promote the development and use of satellite communications and broadcasting services, as well as other aspects of space activities, for the socioeconomic and cultural welfare of the Asia-Pacific region.

El Cluster Latinoamericano de Industria Espacial (CLIE)

El Cluster Latinoamericano de Industria Espacial (CLIE) was formed in 2021. It is composed of Latin American space companies and organizations. CLIE works through collaboration, innovation, and shared resources to promote the region's space industry. Its aim is to enhance space technology development, create market opportunities, and foster regional cooperation.

Commercial Spaceflight Federation (CSF)

The Commercial Spaceflight Federation is a U.S.-based trade organization that mainly focuses on the commercial space transportation industry. CSF was founded in 2006 and, as of May 2024, has more than eighty-five member organizations. The main goals of CSF are to promote technology innovation, guide the expansion of Earth's economic sphere, bolster U.S. leadership in aerospace, and inspire America's next generation of engineers and explorers.

Commercial Smallsat Spectrum Management Association (CSSMA)
The Commercial Smallsat Spectrum Management Association is a U.S.-based trade association that coordinates and advocates for policies affecting small satellite users. CSSMA was formed in 2016 and has over forty members. CSSMA maintains a strong focus on spectrum coordination between commercial users, government users, and satellite operators.

CONFERS (Consortium for Execution of Rendezvous and Servicing Operations)
The Consortium for Execution of Rendezvous and Servicing Operation is an international trade association of private sector entities focused on developing in-space servicing, assembly, and manufacturing (ISAM) capabilities. CONFERS helps to develop international standards and engages with governments on policies and regulations related to ISAM.

European Association of Remote Sensing Companies (EARSC)
The European Association of Remote Sensing Companies is a nonprofit, membership-based organization that promotes the use of Earth observation technology, with an emphasis on European companies that offer Earth observation-related products and services. EARSC's mission is to foster the development of the geo-information service industry in Europe. As of May 2024, EARSC represents 27 countries and has over 140 member organizations. Member of observer status is available to any organization that uses or provides remote sensing observations of the Earth and its environment, irrespective of sensor type or source (e.g., satellite, aircraft, or unmanned aerial vehicle).

The European Space Policy Institute (ESPI)
The European Space Policy Institute was formed in 2003. Composed of European space stakeholders, including agencies, industries, and research organizations, ESPI conducts independent research and policy analysis and hosts events. It aims to provide decision-makers with informed insights to shape space policy and promote Europe's role in space activities.

Global Satellite Operators Organization (GSOA)
The Global Satellite Operators Association is a nonprofit forum of global satellite operators. As of May 2024, it is composed of CEOs from sixty-three member organizations. It aims to be the platform for collaboration between companies involved in the satellite ecosystem globally and a unified voice for the sector. GSOA was previously the European, Middle East, and African Satellite Operators Association (ESOA) and was rebranded as GSOA in 2021.

ANNEX

Satellite Industry Association (SIA)

The Satellite Industry Association is a U.S. trade association representing the commercial satellite industry. SIA was formed in 1995 by several major U.S. satellite companies as a forum to discuss issues and develop industry-wide positions on shared business, regulatory, and policy interests. SIA has established active working groups involved with a host of policy issues including government services, public safety, export control policy, international trade issues, and regulatory issues (satellite licensing, spectrum allocation, and regulatory policy).

The Space Industry Association of Australia (SIAA)

The Space Industry Association of Australia (SIAA) was formed in 1992. It is composed of Australian space industry stakeholders, including companies, research institutions, and government agencies. SIAA works through advocacy, industry collaboration, and events. Its aim is to promote the growth of Australia's space industry and influence national space policy.

SCIENTIFIC, TECHNICAL, AND OTHER ORGANIZATIONS

Scientific and technical organizations are focused on specialized issues or interests within the space sustainability regime. These organizations, which include nonprofit organizations, unions, committees, and more, often serve as facilitators, convenors, and initiators between other organizations in the space community.

The Consultative Committee for Space Data Systems (CCSDS)

The Consultative Committee for Space Data Systems is an international standards body that focuses on communications and data systems standards for spaceflight.

European Organisation for the Exploitation of Meteorological Satellites (EUMETSAT)

EUMETSAT, the European Organisation for the Exploitation of Meteorological Satellites, was formed in 1986. It is composed of thirty European member states. EUMETSAT aims to monitor "weather, climate and the environment from space on behalf of our member states."

The International Academy of Astronautics (IAA)

The International Academy of Astronautics was formed in 1960. It is composed of global experts in astronautics and related fields. IAA works through conferences, publications, and collaborative research. It aims to foster the development of astronautics, recognize achievements, and promote international cooperation in space exploration and technology.

International Astronautical Federation (IAF)

Founded in 1951 by scientists from around the world interested in dialogue and collaboration in the field of space research, the International Astronautical Federation, as of May 2024, has 513 members from 77 countries, including all leading space agencies, companies, societies, associations, museums, and space research institutes worldwide. The IAF holds the yearly International Astronautical Congress (IAC) at a different locale each year, as well as other global conferences on space exploration, space sciences, and related themes.

International Astronomical Union (IAU)

Founded in 1919 and based in Paris, France, the International Astronomical Union is an international scientific organization of professional astronomers focused on astronomical research and education. The IAU is the recognized authority for assigning names and designations to celestial bodies. As of May 2024, the IAU has 85 state members (national bodies) and over 11,500 individual members.

IAU Center for the Protection of Dark and Quiet Skies from Satellite Constellation Interference (IAU CPS)

Within the IAU, the Center for the Protection of Dark and Quiet Skies from Satellite Constellation Interference (IAU CPS) was created in 2022 to coordinate collaborative multidisciplinary international efforts with institutions and individuals. It works internationally to help mitigate the negative impact of satellite constellations on ground-based optical and radio astronomy observations and is also concerned with protecting and preserving humanity's enjoyment of the night sky.

International Amateur Radio Union (IARU)

Founded in 1925, the International Amateur Radio Union is an international union for the cooperation among and coordination of radio frequencies allocated to amateurs, including amateurs using amateur-satellite applications. The IARU has a Satellite Frequency Coordination division, and its IARU Satellite Advisor can help in planning space telemetry, space telecommands, and operating frequencies. Frequency coordination with the IARU is necessary for transmission from space of certain amateur-allocated frequencies.

International Institute of Space Law (IISL)

Founded in 1960, the International Institute of Space Law is composed of institutions and individuals elected based on their contributions to the fields of space law and related social sciences. Dedicated to fostering the development of space law, the IISL organizes and holds an annual Colloquium at the IAC in partnership with the IAF, publishes an annual volume of its proceedings, and organizes an annual space law moot court competition and other events throughout the year. The IISL is a permanent observer at COPUOS, and in recent years has jointly organized a symposium on the first day of the COPUOS LSC meeting.

ANNEX

Open Geospatial Consortium (OCG)
The OGC is a membership-based nonprofit organization dedicated to the development and promulgation of open-source standards for the international geospatial community. Its members include private-sector representatives from the space, airborne, and terrestrial remote sensing industries, government agencies, academia, research organizations, and non-governmental organizations. OGC works through a consensus process to develop standards for the interoperability and sharing of geospatial data, regardless of source. Its membership currently consists of more than 500 organizations worldwide.

ReLACA Espacio
ReLACA Espacio, the Latin American and Caribbean Space Network, was formed in 2021. It is composed of a network of universities and institutions within the region with space-related activities. ReLACA Espacio works through collaboration, knowledge exchange, and joint projects. Its aim is to strengthen regional space capabilities and promote sustainable space development.

Space Frequency Coordination Group (SFCG)
Composed of member agencies including space agencies and international organizations, the Space Frequency Coordination Group works informally to develop resolutions and recommendations which express technical and administrative agreements to prevent and alleviate the risks of radiofrequency interference. The effectiveness of SFCG recommendations depends upon their voluntary acceptance and implementation by members.

Space Generation Advisory Council (SGAC)
The Space Generation Advisory Council (SGAC) was formed in 1999. It is composed of students and young professionals in the space sector worldwide. SGAC works through events and projects to provide input on space policies. It aims to engage the next generation in space policy, advocacy, and international collaboration to shape the future of space exploration.

Space Safety Coalition (SSC)
The Space Safety Coalition (SSC) was formed in 2019. It is composed of satellite operators and space industry stakeholders. SSC works by developing and promoting best practices and guidelines for space operations. It aims to enhance space safety, reduce collision risks, and ensure sustainable use of space environments.

LIST OF ABBREVIATIONS

ADR
Active Debris Removal

AEB
Brazilian Space Agency

AIAA
American Institute
of Aeronautics and
Astronautics

APRSAF
Asia-Pacific Regional Space
Agency Forum

APSCC
Asia-Pacific Satellite
Communications Council

ASI
Italian Space Agency

BELSPO
Belgian Science
Policy Office

BIS
Bureau of Industry and
Security (USA)

BSS
Broadcasting
Satellite Services

CA
Conjunction Assessment

CCL
Commercial Control
List (USA)

CCSDS
Consultative Committee
for Space Data Systems

CD
Conference on
Disarmament (UN)

CDM
Conjunction Data Message

CFR
Code of Federal
Regulations (USA)

CNES
National Center for Space
Studies (France)

CNSA
China National Space
Administration

CONAE
National Commission
on Space Activities
of Argentina

CONIDA
National Aerospace
Research and
Development
Center (Peru)

COPUOS
Committee on the
Peaceful Uses of Outer
Space (UN)

COSPAR
International Committee
on Space Research

COTS
Commercial-Off-the-Shelf

CPS
Center for the Protection
of Dark and Quiet Skies
from Satellite Constellation
Interference

CSA
Canadian Space Agency

CSF
Commercial Spaceflight
Federation

CSpOC
Combined Space
Operations Center (USA)

CTIO
Cerro Tololo Inter-
American Observatory

DAS
Debris Assessment
Software (USA)

DDTC
Directorate of Defense
Trade Controls (USA)

DLR
German Aerospace Center

DoD
Department of
Defense (USA)

DRAMA
Debris Risk Assessment
and Mitigation
Analysis (EU)

DV
Change in Velocity

EAR
Export Administration
Regulations (USA)

EARSC
European Association
of Remote
Sensing Companies

EDAC
Error Detection and
Correction

ESA
European Space Agency

ESG
Environmental, Social and Governance

ESOA
European, Middle East, and Africa Satellite Operators Association

EU
European Union

EUMETSAT
European Organisation for the Exploitation of Meteorological Satellites

EUSST
European Union's Space Surveillance and Tracking

FAA
Federal Aviation Administration (USA)

FCC
Federal Communications Commission (USA)

FMEA
Failure Modes and Effects Analysis

FSOA
French Space Operations Act

FSS
Fixed Satellite Services

GEO
Geostationary Earth Orbit

GEO
Group on Earth Observations

GGE
Group of Governmental Experts

GNSS
Global Navigation Satellite Systems

GPS
Global Positioning System

GSD
Ground Sampling Distance

GSOA
Global Satellite Operators Association

GTO
Geosynchronous Transfer Orbit

HAC
High Accuracy Catalog

HEO
High Earth Orbit

I&T
Integration and Test

IAC
International Astronautical Congress

IADC
Inter-Agency Space Debris Coordination Committee

IAF
International Astronautical Federation

IARU
International Amateur Radio Union

IAU
International Astronomical Union

ICAO
International Civil Aviation Organization

ICG
International Committee on Global Navigation Satellite Systems

ICJ
International Court of Justice

IISL
International Institute of Space Law

ILRS
International Lunar Research Station Cooperation Organization

ILRSCO
International Lunar Research Station Cooperation Organization

IN-SPACe
Indian National Space Promotion Authorization Center

IOT
Internet of Things

ISAM
In-Space Servicing, Assembly and Maintenance

ISO
International Organization for Standardization

ISRO
Indian Space Research Organisation

ITAR
International Traffic in Arms Regulations (USA)

ITU
International Telecommunication Union

ITU-R
International Telecommunication Union Radiocommunication

ITU-T
International Telecommunication Union Telecommunication Standardization Sector

JAXA
Japan Aerospace Exploration Agency

JSpOC
Joint Space Operations Center (USA)

KazCosmos
Ministry for Investment and Development–Aerospace Committee (Kazakhstan)

LBA
Federal Aviation Office (Germany)

LEO
Low Earth Orbit

LSC
Legal Subcommittee (COPUOS, UN)

LTS
Long-Term Sustainability

MASTER
Meteoroid and Space Debris Terrestrial Environment Reference (EU)

MEO
Medium Earth Orbit

MEXT
Ministry of Education, Sports, Culture, Science and Technology (Japan)

MIFR
Master International Frequency Register

MILAMOS
Manual on International Law Applicable to Military Uses of Outer Space

MSIP
Ministry of Science, Information and Communications Technology, and Future Planning (South Korea)

MSS
Mobile Satellite Services

MTCR
Missile Technology Control Regime

NASA
National Aeronautics and Space Administration (USA)

NASB
National Academy of Sciences (Belarus)

NATO
North Atlantic Treaty Organization

NEI
Non-Earth Imaging

NGO
Non-governmental Organization

NGSO
Non-geostationery orbit

NOAA
National Oceanic and Atmospheric Administration (USA)

NOTAMs
Notices to Air Missions

NSAU
National Space Agency of Ukraine

NSC
Norwegian Space Center

NTIA
National Telecommunications and Information Administration

NZSA
New Zealand Space Agency

OD
Orbit Determination

OEWG
Open-Ended Working Group

OFAC
Office of Foreign Assets Control (USA)

Ofcom
Office of Communications (UK)

OGC
Open Geospatial Consortium

OOSA
Office for Outer Space Affairs (UN)

ORSAT
Object Re-entry Survival Analysis Tool (USA)

OST
Outer Space Treaty

PAROS
Prevention of an Arms
Race in Outer Space

Pc
Probability of Collision

PCA
Permanent Court of
Arbitration

PNG
Position,
Navigation, Timing

R&D
Research and Development

RAAN
Right Ascension of the
Ascending Node

REACH
Registration, Evaluation,
Authorisation
and Restriction of
Chemicals (EU)

RF
Radiofrequency

REG
Registration Convention

RHUs
Radioisotope Heat Units

Roscosmos
State Space
Corporation (Russia)

RTGs
Radioisotope
Thermoelectric Generators

SAR
Synthetic Aperture Radar

SDA
Space Data Association

SEEs
Single Event Effects

SEUs
Single Event Upsets

SFCG
Space Frequency
Coordination Group

SIA
Satellite Industry
Association

SLR
Satellite Laser Ranging

SME
Small and Medium Sized

SSA
Space
Situational Awareness

SSC
Space Safety Coalition

SSN
Space Surveillance
Network (USA)

SSOD
Small Satellite
Orbital Deployer

STC
Space Traffic Coordination

STEM
Science, Technology,
Engineering, and
Mathematics

STI
Science, Technology, and
Innovation

STM
Space Traffic Management

STSC
Scientific and Technical
Subcommittee
(COPUOS, UN)

SUPARCO
Pakistan Space and Upper
Atmosphere Research
Commission

SWPC
Space Weather Prediction
Center (NOAA, USA)

TBC
To Be Considered

TBD
To Be Determined

TCBM
Transparency
and Confidence-
Building Measure

TR
Technical Regulation

TRL
Technology
Readiness Level

UAE
United Arab Emirates

UAS
Uninhabited
Aircraft Systems

UK
United Kingdom

UKSA
UK Space Agency

UN
United Nations

UNCITRAL
United Nations
Commission on
International Trade Law

UNESCO
United Nations
Educational, Scientific and
Cultural Organization

UNGA
United Nations
General Assembly

UN-GGIM
United Nations Committee
of Experts on Global
Geospatial Information
Management

UNIDIR
United Nations Institute for
Disarmament Research

UN-SPIDER
United Nations Platform
for Space-based
Information for Disaster
Management and
Emergency Response

US
United States

USGS
United States
Geological Survey

USML
United States
Munitions List

USSR
Union of Soviet
Socialist Republics

USSTRATCOM
United States
Strategic Command

UTC
Coordinated
Universal Time

WMO
World Meteorological
Organization

WRC
World
Radiocommunication
Conference

WTSA
World Telecommunication
Standardization Assembly

INDEX

Made in the USA
Middletown, DE
21 October 2025

19622951R00100